Healthy Soils,
Sick Soils

Healthy Soils, Sick Soils

Understanding the Nature of Humus & How it Functions in Farming Systems

Margareth Sekera
Translated by Paul Lehmann

Acres U.S.A.
Austin, Texas

Healthy Soils, Sick Soils

English edition copyright © 2016
Revised 6th German edition copyright © 2012
OLV Organischer Landbau Verlag Kurt Walter Lau

All rights reserved. No part of this book may be used or reproduced without written permission except in cases of brief quotations embodied in articles and books.

Although the author and publisher have made every effort to ensure that the information in this book was correct at press time, the author and publisher do not assume and hereby disclaim any liability to any party for any loss, damage, or disruption caused by errors or omissions, whether such errors or omissions result from negligence, accident, or any other cause.

Acres U.S.A.
P.O. Box 301209
Austin, Texas 78703 U.S.A.
512-892-4400 • fax 512-892-4448
info@acresusa.com • www.acresusa.com

Printed in the United States of America

Front cover photography © Kesu01/iStock/Thinkstock
Front cover photography © Can Stock Photo Inc./slowbird

Library of Congress Cataloging-in-Publication Data
Sekera, Margareth, author.
    Healthy soils, sick soils : understanding the nature of humus and how it functions in farming systems / Margareth Sekera.
        pages cm
        Gesunder und kranker Boden. eng
    Austin, Texas : Acres U.S.A., [2015]
    Includes index.
    LCCN 2015036680
    ISBN 978-1-60173-126-5 (pbk.)
    ISBN 978-1-60173-127-2 (ebook)
    1. Soils. 2. Humus.

S591 .S43713   2015   631.4—dc23
        2015036680

Table of Contents

Preface..vii

1. Tilth and Soil Structure....................................1
2. Symptoms of Unhealthy Soil..............................20
3. Plant Development in Friable and Non-Friable Fields........36
4. Remedies for Unhealthy Soil..............................43
5. The Soil as a Water Reservoir 57
6. The Spade Test...66
7. Organized Humus Management...........................72
8. Causes of Soil Fatigue 91
9. Is Livestock-Free Agriculture Possible without
   Endangering Soil Fertility?...............................104
10. Fertilization: One Factor in Maintaining Soil Fertility 112
11. Working the Soil in a Biologically Friendly Manner........119

Closing Remarks 135

Afterword..137

Index 139

About the Author 151

Preface

Back when the fifth edition of *Healthy Soils, Sick Soils* was in the planning stages, I asked myself to what extent it should address new ways of using machinery and new agro-technical methods. I ended up largely choosing not to, because this small book isn't intended to give instructions on cultivation, but rather to illustrate connections in the field of soils. Nothing about these relationships has changed, though there have been some new discoveries.

A large number of technical innovations are suggested nowadays, with changes frequently being made and the market constantly being redefined. Which options a farmer chooses depends on his particular circumstances. As far as maintaining soil quality goes, the same basic principle applies now as years ago: any technical innovation must fit the natural requirements of the soil and be as gentle as possible.

Significant changes have taken place in lowland areas in the last decades. Animal-free farming predominates, which entails considerable demands on the soil due to its undiversified economic framework. Only a very few farmers provide their soil with an equivalent of the missing healthy crop rotation and animal manure. Yields are maintained only with very heavy use of fertilizer, irrigation, and very expensive machinery. This is not conducive to productive agriculture. Particularly in cases of extreme strain on the soil, cultivation must be broken up by more extensive crop rotation. Maintaining consistent soil fertility can only be accomplished by working with nature, and never against her, as this is the only way to ensure a successful outcome and that our efforts will not simply be wasted.

I am sincerely grateful for the large amount of interest in questions of soil cultivation.

A special thank you is due to the large number of farmers who have given me suggestions and shared their observations. This book is also dedicated to them, with the hope that it may provide a small contribution to better understanding soil, and that every farmer may observe and know his soil so well that he can serve as its doctor.

Margareth Sekera
Vienna, 1955

Chapter 1

TILTH AND SOIL STRUCTURE

"In good tilth" is an old agricultural term for the healthiest and most fertile possible soil. It refers to soil that is crumbly and well-aerated and doesn't produce any clods when plowed. The essence of being in good tilth is therefore the soil's friability. But a field made friable with a plow or other piece of equipment, by burrowing animals, or by frost might not remain in good tilth for long. It cannot truly be said to be in good tilth until the topsoil has remained friable over an entire growing season and does not break down through the silting effects of water.

When early research was being done, it was already well known that soil quality was very closely linked to crumb structure, but there was a long period of disagreement between soil biologists and colloid chemists over what was responsible for this relationship. The colloid chemists believed that the consistency of the topsoil depended on the water resistance of the soil fabric's binding substances, while the biologists presumed that biological tillage of the topsoil by the microorganisms living in it was the cause. Today we know that both factors are partially responsible for maintaining the topsoil's condition.

All soil-based life, from plant roots to the organisms in the soil, is dependent on its habitat. This habitat consists of a system of hollow spaces made up of many large pores. Depending on their sizes, they fulfill different functions. We differentiate between:

- coarse pores (larger than 0.03 millimeters), which help aerate the soil. Roots and soil organisms breathe, and are therefore dependent on an available supply of oxygen. These air channels in the soil only fill with water temporarily, but they're important for the its rapid distribution through the soil
- medium-sized pores (0.003–0.03 millimeters), which form the soil's water supply system and store rainwater
- fine pores (under 0.003 millimeters), which serve as the last water reserves during periods of drought, helping to preserve life in the soil.

These size-based categories of the pores in the soil are also important to the many organisms of various different sizes living there, which don't always quite coexist in harmony. For example, the bacteria that live in the small cavities are protected from attack by larger organisms, such as protozoa. Any change in the structure of these cavities therefore affects not only the soil's water and air supply but also the distribution of its inhabitants. *Optimal living conditions are produced by the presence of coarse, medium-sized, and fine pores in approximately equal proportions, because this results in a state of equilibrium between aeration, water supply, and water retention.*

The structure and organization of its cavities is an indication of a soil's structure. Optimal conditions are indicated by crumbs of 1–3 millimeters in size (i.e., by a crumb structure with a porous system of cavities). The spaces between the crumbs form a branched network of air channels, while the medium-sized and fine pores are located within the crumbs themselves.

If you use an agricultural implement to break up the soil, the crumbs will settle loosely next to each other and will only touch at single points, making capillary water flow from one crumb to the next impossible. Having them settle in this manner on the field's surface is thus desirable for preventing evaporation, but it would be completely unsuitable around roots because it prevents needed water from being transported through the soil. Mechanically breaking up the soil therefore does not produce the optimal structure, which cannot arise until the field settles to the point that the crumbs grow together and form an interconnected crumb layer. Part of how this takes place is through colloidal chemistry processes, as the flocculated soil colloids form primary aggregates of 0.1–0.2 millimeters via the brick-and-mortar principle. Biological aggregate formation also takes place, in which soil organisms and humic substances form

Figure 1

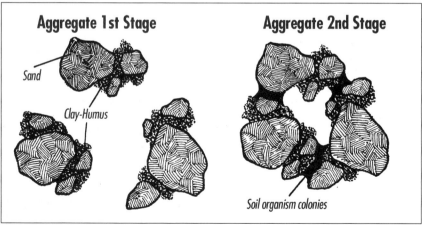

A general depiction of crumb structure.

crumbs of 1–3 millimeters from the primary aggregates (biological tillage). This process is depicted schematically in Figure 1.

This blueprint for soil structure makes it clear that compositional problems can have two different causes. If the primary aggregates are not water resistant due to a lack of colloid flocculation, then the issue is referred to as a loss of structure. If the biological tillage by the microorganisms is inadequate and the crumbs produced by the plow cannot withstand the silting effect of the water, then the issue is referred to as a loss of friability.

A loss of structure is characterized by a swelling and agglutination of the soil pores, causing a disturbance to the sensitive organisms living in them, which prevents stable crumb formation. The resulting compaction of the soil affects both the topsoil and the subsoil, and can be perceived through a reflected light microscope as a waxy, caked-together appearance in its structure. Soil with this sort of structural deficiency requires special remedial measures, and only an experienced soil specialist can give a proper diagnosis and suggest an appropriate remedy. Since this sort of issue is generally rare and only appears in certain regions—and also because this kind of damage cannot be independently fixed by a farmer on his own—we will not address it any further here. In less serious cases, the peptizing and mobile colloids can be flocculated by lime or gypsum fertilization.

A loss of mellowness is caused by low crumb stability due to a lack of biological tillage and humus formation. The compaction of the soil is limited to the topsoil, while the subsoil remains permeable to water and air. The structure appears flaky through a microscope. The photomicrograph in Figure 2 demonstrates the difference between the structures of friable soil and compacted, non-friable soil.

On the left, you can see that the spaces between the crumbs form an interconnected network of coarse pores (air channels), while the

Figure 2

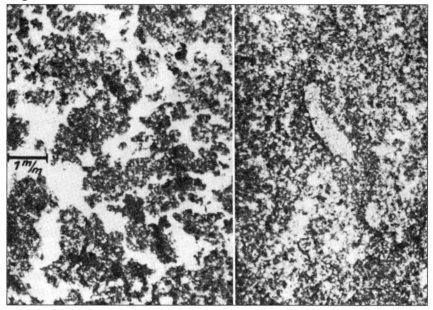

Cross-sections of crumbly (left) and compacted soil.

finer pores lie within the crumbs themselves. In this healthy structure, in which each individual crumb forms its own water reservoir, the rain can quickly fill the cavities which can then slowly supply water over time. A large amount of air can also flow around the crumbs, providing the best possible conditions for organisms living in the soil and for roots.

But now an important question: what's the use in laboriously creating a good crumb structure with a plow and harrow if it's just going to break down again shortly afterward? On the right side of the figure, the image looks so different that it seems to depict completely different soil. This process takes place annually in most fields, however, and the mellowing effect of the plow doesn't last for very long. Usually the deterioration begins so quickly that the topsoil breaks apart and lies in compacted form by early summer. The lack of air channels suffocates all life in the soil. Water circulation and stor-

Figure 3

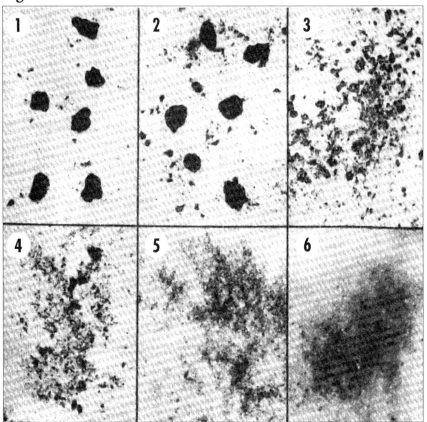

How silty a soil becomes signifies the stability of its crumbs:
1st degree—no breakdown, or only into large pieces
2nd degree—predominantly large, broken pieces, with some small ones
3rd degree—approximately equal numbers of large and small broken pieces
4th degree—predominantly small, broken pieces, with some large ones
5th degree—only small broken pieces
6th degree: complete dissolution of the soil structure, cloudiness in the water

age can no longer take place; the soil quickly becomes too wet and then too dry. Working a field in this condition is very difficult. The plow can no longer properly break up the soil and can only break off coarse clods, requiring expensive further work.

In light of these observations, and in accordance with the need to create stable topsoil, we tested the topsoil's consistency. This was accomplished using very simple methods: 10 grams of air-dried topsoil (2–3 millimeters) was placed in a shallow bowl of water and vigorously shaken. The result was very clearly apparent silt (Figure 3), organized in a series of gradations.

Nowadays there are also quantitative methods for judging the consistency of topsoil, but a layman will be able to achieve sufficient results using just this simple method. You can find a wide range of different conditions from field to field, from good consistency to complete collapse of the topsoil. In order to determine whether the appearance is due to the soil itself or to the cultivation methods used, the immediately adjacent area was also tested using this method. This sort of comparison is always useful and can tell you how different a given field is from the surrounding natural conditions. It was a very insightful comparison that showed us that the untouched balk always displayed better topsoil consistency than the cultivated field.

How significant the differences are can be observed in Figure 4.

These findings proved that the structural decay of a field definitely cannot be a natural aging process of the soil, but rather a malady brought on by cultivation as a result of the agricultural methods used. The less diversified the cultivation methods are, the further the decay progresses. The comparison between the field and the unplowed ridge, or balk, made it clear that it's possible to heal the sick soil of the field, but it's first necessary to determine what was missing in the fields that became silty. What makes the structure of the natural topsoil so firm, and what can cause it to break down?

Next, we took a close look at a droplet of water dropped onto the topsoil. This is one of the most interesting things that you can observe through a simple microscope (approximately 10x magnifica-

Figure 4

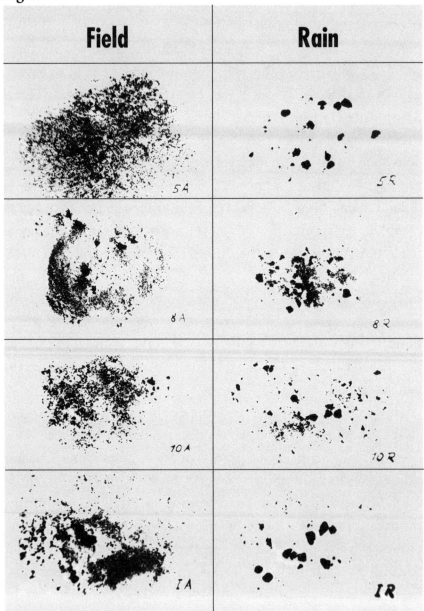

Crumb consistency in a field and its balk.

tion). From looking at the plant roots or microorganisms it appears that a catastrophic earthquake has taken place. First, the water permeates the soil's pores; this takes place very slowly. Then, however, there is an earthquake-like movement through the soil, producing fissures in its structure, sometimes very rapidly and in other cases very slowly. This makes it clear that topsoil is not made up of a pile of individual components, but of aggregates. The aggregates are made up of sand and clay joined like brick and mortar, and many aggregates of this type make up the topsoil. In some soils the crumbs of the topsoil can fall apart upon exposure to water becoming siltier and siltier depending on how small the aggregates are. In other soils fissures can form without the crumbs falling apart. This is true of the firm topsoil of the balk and of healthy fields.

To do research into the consistency of the aggregates, we once again made use of a microscope. This time, we broke up each crumb with a pin under approximately 100x magnification which revealed a loose connection between the aggregates. It was reminiscent of a globe connected by roots. However, this comparison is only valid on a microscopic scale. Much like how a hillside requires a protective cover of vegetation with its root system to avoid being eroded by water, crumbs of topsoil also require similar biological tillage. The same things that plants provide on a large scale are provided here on a small scale by microorganisms—the bacteria, fungi and actinomycetes—several million of which are present in a single gram of soil. It has been calculated that approximately 1,000 kilograms of soil organisms are present per hectare of soil (or 900 pounds per acre) to break down organic substances and allow the transportation of plant nutrients. This means that the weight of the organisms in an acre of soil is the same as the weight of the livestock that can be supported by one acre. The fact that the

Figure 5

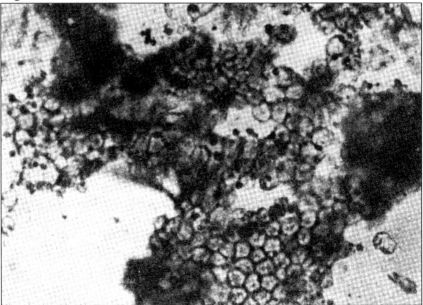

Biological tillage by bacterial colonies.

crumbs are especially densely populated by these microorganisms that proliferate in the soil sometimes was observed by R. Lang and P. Vageler, but the researchers did not draw any conclusions from it. When F. Sekera followed up on the reasons why soil becomes silty in 1938 and observed the presence of microbes in hundreds of soil samples, he suddenly realized the connection between them and the biological tillage forest managers employed on hillsides which were in danger of erosion by planting vegetation that grew as quickly as possible after clearing woodland, limiting the subsequent erosion. This comparison is accurate in every respect; one must simply transfer the concept from the large scale of a hillside to the microscopic scale of the crumbs to recognize that water causes a microerosion in the soil, which can only be counteracted by the organisms in the soil carrying out the proper biological tillage. This new perspective on the nature of soil allowed researchers to over-

Figure 6

Mature biological tillage by azotobacters.

come a roadblock, so to speak, in research on soil conditions and from that point forward, question after question has been successively posed and answered.

The first step was to make the biological tillage in the crumbs visible and to use microphotography to create accurate images of it. Using special methods of fixation and dyeing it was possible to obtain images of the organisms in the soil, a small selection of which are provided after this as a representative picture of the life in the soil.

Figure 5 shows a colony of bacteria (azotobacters). You

Figure 7

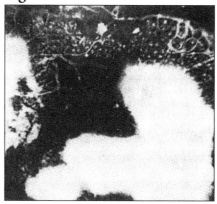

Final state after crumb tillage by azotobacters.

Tilth and Soil Structure

Figure 8

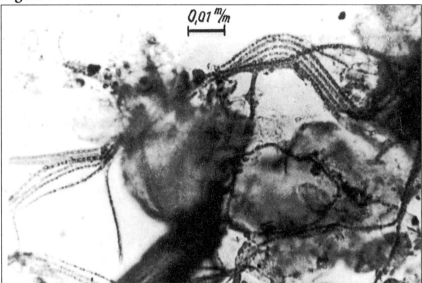

Biological tillage through an organism's strands.

can observe how the cocci grow in even the finest capillaries of the soil. If you follow this settlement's development further, then you can see (Figure 6) how the individual cocci are surrounded by mucus and begin to closely coexist with other organisms, which in this case are filament shaped. In the next stage (Figure 7), the cocci are no longer visible: they've died off. But the mucus they've produced lines the cavities and, in conjunction with the filament-shaped organisms (Figure 8), increases the firmness and flexible cohesion of the soil structure.

The short lifespans of the microorganisms mean that their colonies are always changing in appearance. The decomposition products and residue of the dead colonies serve as new building blocks for the soil and strengthen its connective materials, so this "biological tillage" produces a crumb structure that is more or less stable depending on the activity of the organisms.

Figure 9

Untilled and tilled crumbs in the rain.

The second step in researching this phenomenon was to prove that this biological tillage actually produces improved stability in the topsoil and can avert the damaging effects of water. Today, we have research methods that can control the biological tillage and its fluctuations. A particularly demonstrative example is an instructional experiment in which crumb consistency is tested using only the simplest methods. As shown in Figure 9, crumb samples are exposed to rain (sprayed with water) on a sloped surface. The only difference between the preparations of the two samples is that the sample on the right was admixed with approximately 0.5 percent organic substances (e.g., alfalfa meal), after which both samples were kept moist for two to three weeks so that the microorganisms had enough time to reproduce based on the available nutrients and to reinforce the crumbs.

As the experiment makes very clear, the "unreinforced" crumbs were turned to silt by rain, whereas the "reinforced" ones maintained their structure. (If you add a bacteria poison into the experiment, then the crumbs do turn to silt despite the added organic substances; this serves as evidence that it's not the organic substances themselves, but rather the microorganisms which they feed that make the topsoil stable.)

During World War II American soil scientists also thoroughly examined this effect of the biological formation of aggregates in the soil. The term "biological tillage of the topsoil" has become a staple of scientific and practical agricultural literature, and we can give the term "in good tilth" a clear definition:

Being in good tilth refers to soil that has a crumb structure formed by the biological tillage of the microorganisms in the soil.

When a farmer says that his field is "in good tilth" after working it with a plow, he means that that topsoil created by the plow has been "reinforced" by the soil microbes, which maintains its consistency, which is a requirement for optimal activity in the soil to continue over time. On the other hand, silt formation and compaction are signs that sufficient biological tillage has not taken place. The plow may have physically crumbled the soil, but the microorganisms failed to produce good tilth.

The question is now what the farmer must do to facilitate biological tillage and produce good tilth in a plowed field.

Three conditions must be fulfilled:

1. The primary aggregates must be water resistant, because colloidal microscopic and submicroscopic components cannot be bound together by the microorganisms. The larger and bulkier the aggregates are, the easier they are to reinforce. If the soil colloids are present in a peptized and mobile form, the lime state must be corrected by liming with chalk or gypsum and colloid flocculation must be induced. Dissolution of the soil colloids leads to a breakdown of the structure (i.e., to a dissolution of the aggregates or to swelling and cementing of the pores), which inhibits the proliferation of organisms.

2. The soil organisms must be supplied with enough organic nutrients. These are provided by the mass of roots produced in the soil, by plant parts on the surface, and by organic fertilization. If the organisms do not have sufficient nutrients available, the result will be poor biological tillage and low soil quality.
3. The soil climate must be balanced. The goal must be for the soil to always be covered by vegetation or by a layer of plant litter. This protects the biological tillage processes in the topsoil from rainfall, dehydration and drift.

All of the practical methods of making soil more friable are based on these three fundamentals. Nowadays, the traditional understanding of friability brought on by frost or working the soil via mechanical methods needs to be amended. An agricultural implement or frost can only mechanically crumble the soil; they can never put it in good tilth. It is therefore inaccurate to say that soil is "in good tilth" after the winter's frosts have made it friable and produced a satisfactory seedbed through purely physical processes. The decisive factor is whether biological tillage takes place after this physical crumbling of the soil and whether the organisms in the soil make the topsoil stable. Plants are the first step toward good tilth because underground their roots provide nutrients for the organisms in the soil and above ground they shade the soil and protect the mellowing processes against disruptive influences. Because of this friability is dependent on the life cycle of the plant population. Due to the short lifespans of the microorganisms, conditions are never static, but rather continuously being built up and broken down as long as the bacteria's food accumulates and the soil has a protective plant layer.

The more organic matter is generated in the soil, the better the biological tillage works. Periods without vegetation mean times of starvation for the organisms in the soil and thus lead to the biological tillage slowing down and an increased risk of silt formation.

But even the organic connective material and mucus masses of the soil organisms aren't enough to provide lasting protection against silt formation if the food available to them has been fully used up and consumed into carbon dioxide. Sustained protection against silt is only possible if humus is also formed during the breakdown of the organic matter and cements the soil particles together or if the walls of the cavities are coated in it. Microbes (bacteria, fungi, etc.) can only form humus, however, in concert with animal organisms (worms, mites, springtails, etc.). Because of this, it's important to supply the soil with a community of microflora and microfauna. Until this is done the primary aggregates will not receive the necessary biological tillage and humus cementing to guarantee lasting consistency.

This cooperation between higher plants, microflora and microfauna is known as "biocenosis." All of the participants are equally dependent on each other. The plant cover provides the food for the soil organisms and creates the favorable soil conditions necessary for them to function. The microbes and small organisms in the soil, for their part, create an optimal habitat for the plant roots.

No field remains in a static condition after being broken up. There are always two processes working against each other: loosening and compacting. Human efforts serve as the countermeasure to the silting effects of water. It's not just when the soil gets wet that it becomes more compacted, but also when it dries out. In both cases, water is in motion and microerosion is carrying out its destructive effects. The soil structure is therefore governed by constant conversions that take place as follows:

The arrows should indicate that constant motion is the rule in soil structure. Arrow 1 represents the soil-mellowing processes (working it physically, biological loosening, frost). Arrow 2 represents the processes working against them to compact the soil. The mellowing initially produces an unstable crumb structure, which is constantly liable to soil compaction and can therefore only produce a temporary improvement unless biological tillage (arrow 3) is taking place, producing a stable soil structure. The soil will maintain this condition for as long as living conditions are favorable to the microorganisms and they can produce humus. If they can't, the soil will become less friable (arrow 4) and will again be compacted.

This new information about the life in the soil must now be integrated into our existing knowledge of soil biology. The core of it is: soil and plants form a biological unit; plants both give to and take from the soil. Their roots give organisms living in the soil their basic nutrients—soil without plants would be barren. This fundamental truth is hard to recognize if you study the processes of soil formation, and plant nutrition and the laws of water and air balance separately. People notice these individual issues in isolation from each other and can easily fail to notice the natural relationships. But the issues are intertwined like the gears of a clock, resulting in the interrelationships depicted in figure 10.

This outline is based on the fact that the amount of root matter present in the soil determines the amount of nutrients available to the organisms living in it, thereby determining the organisms' productivity. It's best to think of the outline as a lengthy illustration of the relationships between the living organisms, soil respiration,

Figure 10

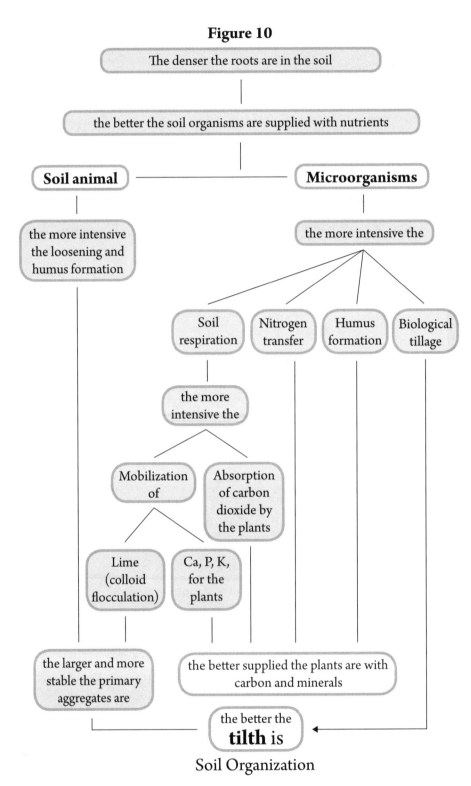

Soil Organization

plant nutrients, and humus and good tilth formation. The natural law governing a field reveals itself there in its simplicity and utility.

It is:

The denser the roots in the soil, the better the nutrients available to the plants and the more complete the humus and good tilth formation will be.

Any impediments to root development will thus lead to a lack of friability. This natural law must be respected whenever you start to work a field. It follows a fundamental of law of life in general, namely the principle of mutual cause and effect. In this case, this means that increased root production leads to more favorable structural conditions in the soil, which then leads to increased plant growth. And the opposite is also true: limited root production causes the soil structure to deteriorate, which then leads to even more limited plant growth. A one-time degradation in soil quality will snowball inexorably forward until its causes are identified and addressed.

This outline also demonstrates the key role that soil quality plays in nature. The consistency of the crumb structure determines how long plants can enjoy the optimal growing conditions that result from tilling a field. It also determines whether the soil can function effectively as a nutrient and water store, and there is practically no aspect of cultivation that shouldn't be considered from the perspective of soil quality. K. von Rümker once put it very aptly: the production and maintenance of good tilth is the key to successful cultivation.

Chapter 2

SYMPTOMS OF UNHEALTHY SOIL

Structural deterioration takes place to a greater or lesser degree in almost every field, and its consequences include soil that is too silty or too compacted. To make the steps that take place in this situation clear, the sequence of images in Figure 11 depict the inflow and draining of the water and the resulting changes sustained by the soil.

The images illustrate the water permeating the soil and the increasing saturation that results. The image in the upper left clearly shows the porous, spongelike structure of crumbled soil. The coarse pores are still filled with air, but the crumbs themselves are already saturated with water. Scattered air bubbles are contained within the interiors of the crumbs.

The image in the upper right shows water flowing out of the saturated crumbs and coating them with a thin film of water. Since the transfer of water from the fine pores into the coarse pores produces volatile changes in its surface tension, this process also causes a physical attack against the soil's structure. While flowing out, the water tears off pieces of soil from the crumbs, beginning the process of microerosion.

In the image in the lower left, the water saturation has proceeded further. Instead of interconnected air channels, there are now large pockets of air inside of the coarse pores. The fine material that has collected now flows through the soil in the films of water.

The image in the lower right depicts a state of full water saturation is depicted in. Other than some small embedded air bubbles, the entire volume of the cavities is filled with water. In the upper area, the structure has already broken down. Comparing the four images provides a simple demonstration of the destructive effects of water.

Even more blatant structural changes are brought on by the outflow of the water. Air begins to flow into the coarse pores once again. The air bubbles that are embedded in the channels move closer to each other. This takes place via pulsating, fitful movements. It causes some of the channels to widen, while other parts of the cavities fill gradually. In the remaining channels the water flows in a film-like manner along the sides and carries eroded material along with it. The fitful forward motion of the water-air menisci especially impacts areas where eroded material is stored. The movement of the water in the cavities is not consistent; you can frequently observe water moving along with eroded material on one side while causing the soil to become silty on the other. The effects are no different than when a flowing river carries away material along one of its bends only to deposit it further along its course.

In its final stage, the water is so thoroughly drained that only the walls of the large cavities are still covered with a film and no further water movement is easily perceptible, though we can assume that some is still taking place. The dense compaction of the soil can be clearly seen at this point.

Figure 11

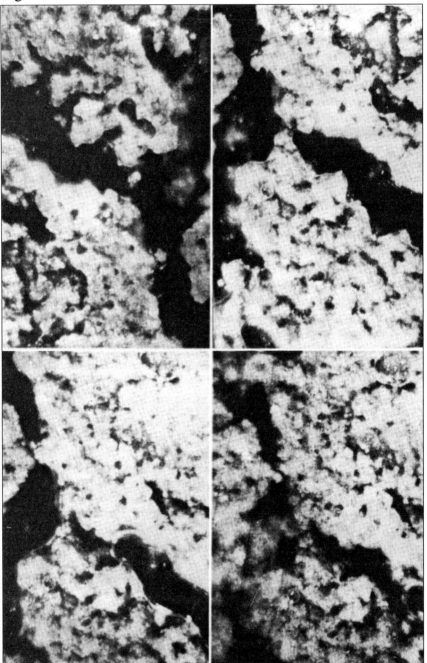

Crumb structure microerosion when moisture penetrates the soil.

Unfortunately, it isn't possible to make out the motion of the water and the resultant transfer of the soil material from these photomicrographs. Since the flowing water film transports fine eroded material, the direction of the flow is visible, however it frequently switches, especially when water escapes. It's common for one part of a channel to be coated with flowing water while the other part is covered by apparently motionless water. The violent movement, especially while water is flowing out, often breaks off whole chunks of soil and washes them away. It's not unusual for this process to also cause old channels to fill up and new ones to form. If you can visualize the idea of this happening several times over the course of a growing season, then you'll understand why the soil can be so compacted at harvest time. As this compaction increases over time, however, the water's flow rate will become more sluggish, mitigating the microerosive effects.

The less stable the crumb structure, the more extreme these effects will be. The destructive power of the water decreases as the stability of the crumbs increase through biological tillage. In freshly worked or fallow soil, water can attack much more furiously than in soil in which living organisms have formed a solid structure through biological tillage and the formation of a humus lining.

Microerosion, however, is only the first stage of a much more irritating obstacle, namely, macroerosion. When you think of erosion, you usually think of the vast areas of American farmland so threatened by annual soil losses that it took an extensive aid program to provide them with relief. Today, the Natural Resources Conservation Service (NRCS), formerly the Soil Conservation Service, encompasses practically all of the farmland in the United States, and it's not an exaggeration to say that almost all cultivation in America is characterized by the struggle against erosion. This danger is not as great in central Europe because of the differing climate conditions;

North America is afflicted by significantly heavier rainfalls, which have a much more destructive effect on soil structure than the rain in Europe. But it would be very unwise to discount the erosion problem in Europe as a secondary concern. The danger exists in Europe as it does in North America and requires preventative measures.

Any slope, even the smallest depression, carries the risk of soil erosion due to downward-flowing water. We differentiate between two different forms of erosion: sheet erosion and furrow erosion. The latter takes place in furrows in the soil, where water collects and carves deep furrows as it flows, making the destructive effect of water obvious to anyone. Less conspicuous but more common and significant is the damage caused by sheet erosion, in which water washes away fine soil from the surface and deposits it into depressions in the earth. In flat or gently rolling terrain, this causes the formation of the well-known "loam crests," which always cause problems with working the land and are responsible for erratic crop growth. Sheet erosion is especially perceptible when the sterile subsoil becomes visible. It's common to find different soil compositions in a small area without ever considering that sneaking sheet erosion is taking place, a constant potential threat to a farmer's work.

A more in-depth look at the problem tells us the following: the primary cause of erosion is absolutely not the downward-flowing water, but rather the fact that the field is not absorbing the water quickly enough. Friable soil with a structure that hasn't been broken down by rain and has a gradual transition between the topsoil and the subsoil will certainly absorb rain faster than topsoil that breaks down in the rain and accumulates such a backlog of water that it can only flow away via the surface. The subsoil can absorb water many times more quickly if there's no layer of compacted topsoil acting as a barrier. This is thus the primary cause of erosion. It begins with the "microerosion" in the soil, which causes the individual crumbs

Figure 12

Sugar beets exposed by erosion.

to lose their water resistance and to dissolve in the rain. "Macro-erosion" first sets in when water can no longer be absorbed quickly enough or properly distributed due to structural breakdown. The more fundamental cause of soil erosion is therefore a lack of friability in the soil, and both can be considered maladies of a cultivated field.

Figure 12 shows a beet plot that has been affected by sheet erosion. The beets are fully exposed and the soil is so crusted that it will have to be plowed over and tilled anew.

With this in mind it's possible to take a symptomatic approach to fighting erosion (i.e., to remove the appearance of erosion by minimizing how much water drains off of the slope). Plowing across the slope, making use of grass balks, and building terraces are all strategies that can help as they restrict the flow of water and in doing so help ensure that the fine earth is redeposited. Instead of these methods of defensive warfare against erosion, however, it seems more promising to attack the root of the issue and to eliminate what's causing the damage—in other words, to take an offensive approach. This can be accomplished by increasing the stability of the tilled soil and above all by making sure that transfer between the topsoil and the subsoil remains possible so that the field can quickly absorb water. Due to their heavy rains, Americans must make use of every available method of erosion resistance, fighting the erosion both offensively and defensively. In Europe, a prevention-focused approach is possible, and it seems preferable to not just combat the visible effects of erosion but to eliminate the causes as well. Any regimen of soil care must also encompass this task, and with its help it's possible to master soil erosion.

Thus far we've only mentioned erosion caused by water, but wind erosion is no less dangerous. However, it doesn't generally affect as large an area and thus can be looked at as a local phenomenon. But this sort of damage becomes more likely and poses a constant threat to affected areas in the spring, once the frozen soil has broken down into a loose powder and is scattered by wind. Wind blowing the soil away represents two different dangers to young plant populations. The first is that they can be exposed if the topsoil blows away or be covered by foreign soil blown onto the field. The second danger, however, is not posed by soil in the air, but by the frequently sharp-edged aggregates that roll across the surface of the fields which can physically damage the young plants. Laboratory experiments have shown that only soil that isn't friable can undergo the undesirable buildup of powder that leads to drift. With enough biological tillage and humus formation in the topsoil this powder effect will be averted, meaning that it is very possible to employ soil care techniques against the damage caused by wind erosion.

We presented the microscopic images of structural breakdown in order to instill a better understanding of the process, but how does this information become clear to a farmer? J. Görbing was the first to successfully use a spade test to make this sort of breakdown in a field visible and to reach a diagnosis from the symptoms present. We will go into his methods later; first we will examine the changes in a cross-section of the topsoil when there is insufficient tilth and humus formation from the point when the field is plowed until the harvest in a sequence of schematic images (Figure 13).

Image A shows freshly plowed soil and the unworked subsoil underneath. A furrow depth of 10 inches (25 centimeters) was used. What will happen if the resulting crumb structure isn't stable enough? The silting effect of water takes hold in two locations (Image B). The pounding rain has caused surface compaction (Sc) as it hits and silts the crumbs. The crumb structure has also been attacked at the border of the worked area, however. Since the water

Figure 13

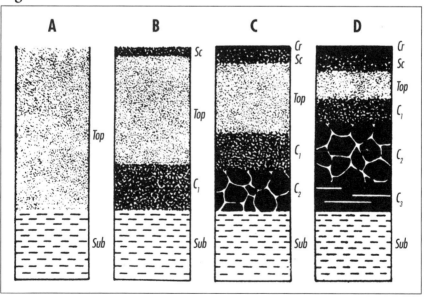

Structural breakdown in the topsoil.
Top = topsoil; Sub = subsoil; Cr = Crust formation; Sc = Surface compaction; C_1, C_2, C_3 = Compaction levels 1, 2, 3.

permeabilities of topsoil and subsoil are different, it's easy for water to become congested in this area. The macerated crumbs break down under the burden of the topsoil above them, first creating some areas of increased compaction, which subsequently cause a corresponding compression of the topsoil (C_1). But it's not just the thickness of the compacted area, but also the degree of compaction that increases (C and D). Although it was originally crumbly and made up of small clods, with increasing and decreasing moisture the clods became coarser (C_2) and plate-like (C_3). By the end, only a 2-inch-deep (5-centimeter) crumbled portion remains of the original 10-inch (25-centimeter) furrow. The individual stages of the compaction are built up level by level from the bottom of the furrow, meaning that the lowest levels are under the greatest strain and it is there that the compaction is furthest along in its progression.

The rapid drying on the surface encourages crust formation (Cr), which threatens the air exchange and thus also the respiratory pro-

Figure 14

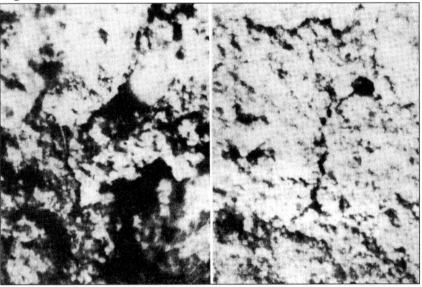

The subsoil (left) is more porous than compacted topsoil (right).

cesses in the soil. This leads to both water congestion on the surface and unhelpful losses through evaporation. The increasing compaction of the topsoil, starting from the bottom of the furrow and moving upward, initially blocks the water and air circulation between the topsoil and the subsoil. As the compaction increases further, the topsoil becomes increasingly waterlogged, but this water supply is unlikely to be transferred to the reserves in the subsoil when it's dry again. This sort of field is particularly liable to both excessive moisture and to drought.

By harvest time, the plowed topsoil is thus considerably denser than the unworked subsoil. You can see from the photomicrographs (Figure 14) how porous the unworked subsoil and how densely silted the topsoil of the very same soil can be.

There have been hundreds of cases where the progression of a field's structural breakdown has been tracked and its maladies recorded. Figure 15 uses some deeply plowed beet plots to show how the topsoil can break down. As the surface compaction is repeatedly broken up, the compaction of the topsoil in the interior of the field increases continuously, sometimes more, sometimes less. In serious cases it can reach the point where all of the topsoil is compacted together.

Image A: 12-inch-deep (30-centimeter) plowing results in satisfactory topsoil. Its deterioration has already begun in December. As surface compaction takes place, which will later be broken up again by plowing but then happen again, areas of increased compaction form in the lower part of the topsoil, which end up reaching C_2 level. The compacted area of the topsoil becomes larger and larger, and by September it has expanded to the 7- to 12-inch- (20- to 30-centimeter-) deep layer. Of the original 12-inch-deep (30-centimeter) topsoil, only the area from 1.6–7.8 inches-deep (4–20 centimeters) is still friable.

Image B: Essentially the same process, but with the breakdown of the topsoil not beginning until April.

Image C: The cross-section of the topsoil remains unchanged until the end of June, but then breaks down very quickly so that only 5 inches (13 centimeters) of the original 12-inch-deep (30-centimeter) plowed topsoil is still friable.

Image D: Structural breakdown already setting in by fall. The original areas of higher compaction reach C_1 level by January, which in April proceeds to C_2. The latter expands constantly in the early summer, uniting with the surface compaction in August, the final result being topsoil that is compacted throughout.

The difference lies in when the structural breakdown sets in. This timing is crucial in determining future yields because the later it sets in, the more can be harvested in spite of the emerging compaction. The sugar beet plots from Figure 15 produced the following yields:

Image A: 9.7 tons per acre (220 decitons per hectare)
Image B: 13.4 tons per acre (305 decitons per hectare)
Image C: 15.2 tons per acre (346 decitons per hectare)
Image D: 9.5 tons per acre (215 decitons per hectare)

If the topsoil is already becoming compacted in winter or spring, water congestion causes the topsoil to dry out more slowly, which disrupts the root systems of the young plants from the very beginning. If the compaction doesn't take place until the growing season so that the root systems are undisturbed, then at worst only the secondary growth of the beets will be affected. Delaying the onset of structural breakdown therefore represents a significant step forward.

Figure 16 shows the progression of the malady in cereal crops. This case is somewhat different to the extent that the less deep seed

Figure 15

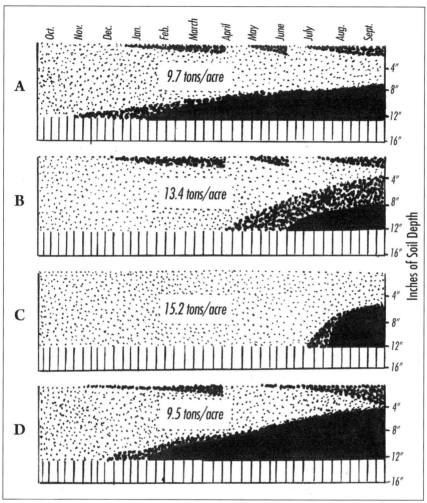

Structural breakdown in deeply-plowed beet fields.

furrows generally do not affect the lower segment of the topsoil, which formed underneath the previous crops. This represents a significant disadvantage as it means that the cereal crop topsoil is separated from the water reservoirs of the subsoil by a barrier layer from the very beginning.

Image A: The 8-inch-deep (20-centimeter-deep) seed furrow does not disturb the remnants of last year's C_2 area from 8–12 inch-

Figure 16

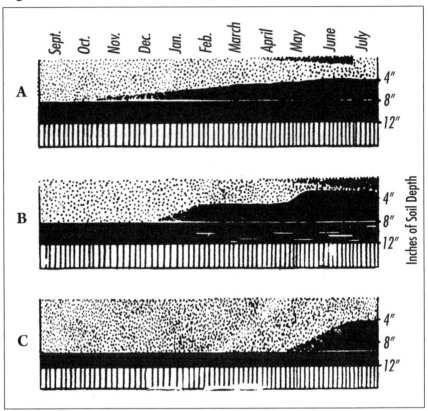

Structural breakdown in cereal plots.

es (20–30 centimeters), so the new topsoil has no connection to the subsoil. The topsoil is already becoming more compacted by the beginning of November, and the process carries on continuously such that the compacted area reaches from 4–12 inches (10–30 centimeters) by May. Of this the layer from 4–8 inches (10–20 centimeters) is newly formed while the layer from 8–12 inches (20–30 centimeters) is left over from the previous year.

Image B: The structural breakdown sets in later, but thus all the more severely. By June, the surface compaction and the topsoil compaction have formed one continuous mass. Toward the end of the winter, the remnants of the C_2 area from the previous year reached C_3 level.

Image C: The cross-section of the topsoil remained undisturbed until the end of June. Then a C_1 area suddenly appeared in the 4- to 9-inch (10- to 23-centimeter layer, quickly transitioning to C_2 level.

When the structural breakdown of the topsoil sets in, how quickly it then proceeds depends on a number of factors. Prolonged periods of rain in fall (i.e., at a time when the topsoil is not yet fully formed), as with seed furrows made in late fall or even winter, make an early breakdown more likely. Poor living conditions for the microorganisms, either an excessively limited supply of nutrients or a lack of a protective ground cover, inhibit their activity and cause a structural deterioration.

There's no way to influence rainfall, but you can change when you do your plowing and the conditions for the soil organisms. Since plants always shed parts of their roots, you can expect a mellowing aftereffect from the preceding crop and a mellowing effect from the current crop. If not enough root matter is shed, or if the time gap between the harvest of the first crop and the growth of the second crop is too great, then the microorganisms will not have enough nutrients available, resulting in the aforementioned effects.

Sowing legumes along with cereal crops is known to be very effective. For this reason, we examined structural deterioration in barley crops with and without the addition of clover seeds (Figure 17).

Image A (left): Without the addition of clover seeds. The 8-inch-deep (20-centimeter-deep) seed furrow did not disturb the remnants of the previous year's sugar beet crop in the 8- to 10-inch (20- to 25-centimeter) layer underneath. By April compaction had set in again and rapidly moved upward in May so that only the top 3 inches (8 centimeters) were still friable by harvest time.

Figure 17

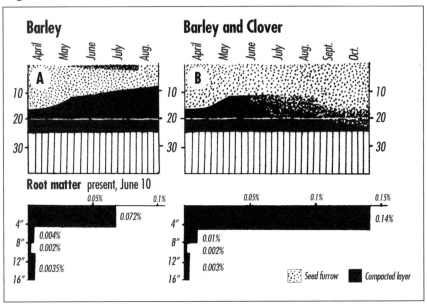

Planting clover loosens compacted topsoil.

Image B (right): With the addition of clover seeds. Until mid-June, the breakdown proceeded on the same scale as in Image A. In July, however, it was clearly noticeable that the preexisting compacted topsoil was wearing away; it disintegrated into smaller individual areas of compaction. At harvest time, a 6- to 7-inch (15- to 17-centimeter) friable upper topsoil layer was present, significantly more than with the unmixed barley. Over the course of the following months, the clover roots broke up the lower soil layers even more, even reaching into the compacted layer from the previous year. In November, water and air exchange with the subsoil was possible once again.

Investigating the root content of the individual topsoil layers revealed two things. First, it revealed how sensitively the roots react to soil compaction, and how rapidly root content drops in the topsoil as compaction increases. Second, it made the increase in root production caused by adding the clover seeds visible.

The clover brings major tilth formation which persists through the next growing season if the plot is closed off enough. An even faster option than clover, which takes longer to take effect due to its slow juvenile growth, is the rapid-growing sweet pea. An experiment with a plot of just rye and a plot of both rye and sweet peas showed new C_2 formation in the 8-inch-deep (20-centimeter-deep) seed furrows of the unmixed rye, which then steadily expanded. In the mixed plot of rye and sweet peas, on the other hand, all that formed was a C_1 layer barely 1.5 inches (4 centimeters) thick. We should also note that surface compaction appeared in the unmixed plot starting in June, but did not happen at all in the mixed plot. At harvest time the unmixed rye had an active topsoil depth of barely 4 inches (10 centimeters) available, whereas the mixture had 6.7 inches (17 centimeters) of friable soil available. The washing away of root matter at the end of May caused an erratic drop in the amount of roots in the compacted topsoil. Adding sweet pea seeds increased the root content from 0.14 to 0.22 percent. We can thus conclude that rye roots receive significant benefits from the presence of neighboring sweet peas.

Essentially, unhealthy soil is characterized by a progressive breakdown in the plowed topsoil. Despite employing many different efforts to create an optimal seedbed in order to make it easier for the roots to penetrate into the subsoil, farmers are disappointed every year when a barrier layer forms between the topsoil and the subsoil during the growing season continually reducing the friability of the topsoil. Because plants' greatest needs are nutrients and water, their root area is more or less strongly constrained.

Chapter 3

PLANT DEVELOPMENT IN FRIABLE AND NON-FRIABLE FIELDS

Any measure taken to improve a field, be it plowing, fertilization, crop rotation, etc., is employed with the goal of creating a healthy plot that will provide a good yield. But if the soil structure loses its consistency, all of these other measures will be for naught.

Compacted topsoil seriously limits the potential development of plants. The prevailing lack of air is already enough to inhibit their development. The active root area is limited to the portions of the topsoil that are still friable. Plants can no longer utilize any sources of nutrients that lie within the compacted area of the topsoil, and increasing applications of fertilizer are necessary to satisfy their nutritional needs. Most important of all, however, is that the plant is cut off from the water reserves in the subsoil by the emerging barrier layer and left vulnerable to summer droughts. Additionally, each heavy rainfall causes a buildup of water above the compacted topsoil, which causes the topsoil to become waterlogged. The plants thus alternate between too wet and too dry, and this imbalance in their water supply weakens their natural resistances so much that they become especially sensitive to disease or pests.

How strongly a plant reacts to these structural disturbances can be determined from its roots. The plant speaks! Its roots are a reflection of the soil and reveal any problems caused by it. This is important to the practical farmer because the plant roots provide clear evidence of any missteps made during tilling and of the condition of the soil.

With cereal crops, surface compaction and depth primarily determine the form of the root network. Friable soil generally leads to a regular system of radicles, crown roots and adventitious roots. The roots can then spread over larger areas, forming a clump-shaped mass. If the soil starts out friable and then later becomes compacted on the surface, then the radicles and crown roots will spread normally, though the adventitious roots will push steeply downward. If the plant's seedbed is compacted from the start, then all of the roots will extend downward vertically. For cereal crops, the way in which the roots form is very significant. While outward expanding roots also permeate the area between the rows and can make use of that soil's nutrient stores, steep vertical root systems cannot reach these areas. This leads to both wasted fertilizer and to an intensification of the compaction of the soil, since a considerable portion of the topsoil enters a new growing period without bacteria nutrients, jeopardizing the biological tillage and the formation of new humus.

How do plants with taproot systems react to structural issues? A particularly demonstrative example is the root system of the sugar beet because both its longitudinal growth and its secondary growth adjust to the structure of the topsoil. Even by the time the sugar beets reach a sugar refinery, they can still reveal the condi-

Figure 18

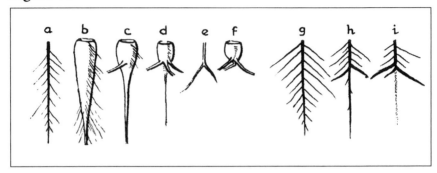

The influence of topsoil compaction on taproot plants.

tions of the supplier's field. Figure 18 shows the various ways that the sugar beet can develop.

a. In undisturbed topsoil, a straight taproot with numerous lateral roots develops.

b. If the structure of the topsoil remains undisturbed, then the secondary growth will lead to the formation of a spindle-shaped beet body from whose root channels many small lateral roots will emerge in all directions. The area around the tip is especially densely filled with rootlets. This optimal form is a sign of a friable beet field.

c. If a thick compacted layer is present in the topsoil, secondary growth will be restricted in that area (shallow beet). The point where a lateral root becomes stronger is usually a very reliable indicator of the depth the blockage begins at.

d. If there is a very serious lack of air in the compacted topsoil, then the taproot will eventually die. Its functions will be taken over by some of the lateral roots with which the beet sits on the compacted area. This sort of "rooted beet" can also arise from an initially entirely normal beet setup that then runs into compaction as it develops.

e. This young beet root has already been disrupted by compacted topsoil in its early stages; it branches at the point where the obstruction begins.
f. Later on, a "rooted" beet will develop from this disrupted root system.

The other plants with taproot systems (rapeseed, horse gram, lupine, etc.) react similarly to the structural conditions in the soil and images g, h, and i depict the forms that can occur.
g. In healthy soil an elongated taproot forms with lateral roots that gradually taper off as it goes deeper.
h. In cases of moderate compaction of the lower topsoil the taproot still remains intact. However, the lateral roots quickly die off and one or two strengthened lateral roots indicate the depth at which the obstruction begins.
i. In cases of serious compaction of the lower topsoil the taproot dies and the plant sits on the obstruction with a few strengthened lateral roots.

The images of cross-sections of exposed rapeseed and potato topsoil (Figures 19 and 20) provide further substantiation of the schematic drawings.

In its juvenile stage, the rapeseed root was still able to push downward unhindered, but the topsoil compaction, which happened later, killed off the lateral roots.

The root system of the potato is a very instructive example. The roots and bulbs are limited to the friable upper topsoil. The compacted area of the topsoil does not help in production, instead hindering it by not allowing any water flow.

Any farmer can observe all of these things in his own fields and learn a great deal from them. All you need is a spade to take samples

Figure 19   **Figure 20**

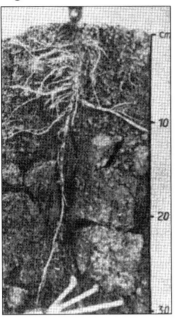

Disruption to the topsoil, disruption to the rapeseed root.

Potato ridge: roots and bulbs limited to the loose upper topsoil.

of the soil here and there. **A spade turns a farmer into the soil's personal doctor**, since what you can see and learn from a simple spadeful of soil is often more than what you can achieve through complex laboratory experiments. You can see a cross-section of the topsoil and the roots growing within it—and the roots reflect the soil structure.

Almost every farmer finds something worthwhile in this sort of spade sample from his fields, and many have claimed that using this method was the first time they were really able to understand their soil rather than simply being reliant on fate. Figure 21 shows how seriously breakdowns in soil structure must be taken, as it demonstrates the relationship between how friable the soil is and the yield produced at harvest time.

Figure 21

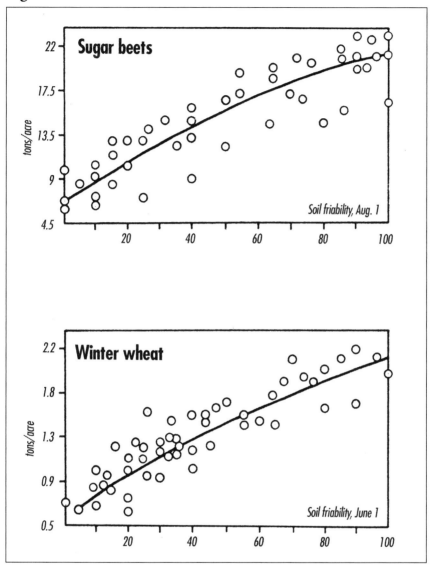

Friability and yields.

Keeping in mind past experience that the crucial period for soil conditions in determining sugar beet yields is in August and the crucial period for wheat is in May and June, we recorded our results and drew a curve showing average yields. Some might question this sort of statistical method since yields are certainly not solely dependent on the structural conditions of the soil. The considerable variance between the individual values speaks to this, but the mere fact that despite differing applications of fertilizer, differing amounts of rain, and various different maintenance techniques that a clear relationship between the harvest and the structural conditions of the soil is still evident proves their key role. The yield curve also proves that a field's average yields can be increased even further by ameliorating the compacted soil and carefully avoiding it in the future. And this doesn't only apply to certain soils. Soft soils can be afflicted by soil compaction just like hard ones. It's true that the breakdown would take place more quickly in some fields and more slowly in others, but in any case it's a symptom that progresses and intensifies from year to year. And this is also true of its remedies. Some soils react the same way to remedial measures, while others require extensive efforts to eradicate the accrued damage. There are many methods involving increased amounts of effort and fertilization that can hold back the damages and produce high yields, but these yields are achieved at the cost of productivity, which is certainly not as desirable as using organized methods of soil management.

Chapter 4

REMEDIES FOR UNHEALTHY SOIL

The preceding chapters demonstrated that the structure of non-friable soil exhibits poor stability and is caused by insufficient biological tillage and humus formation. Soil that is no longer friable is subject to becoming excessively silty or compacted. Remedying this issue involves accomplishing two things:

1. The disruptive compacted topsoil must be physically broken up in order to open up a connection to the subsoil.
2. The newly loosened soil must be filled with new life to get to the point of friability. This means furnishing it with "bacteria food" so that biological tillage and humus formation can take place, maintaining the crumb structure produced by the plow.

Dealing with soil that has lost its mellowness therefore requires both a technical and a biological element. The first part can be resolved through a thorough plowing. The second part can only be accomplished by the plant, by rapidly and densely expanding its root network through the newly loosened soil.

Traditionally, people have only considered plant roots in terms of their functions as anchors in the soil and to absorb nutrients and transfer them to the aboveground portions of the plant. This ignores the fact that the root matter that accrues each year serves as nutrients for the organisms living in the soil. It was thought that a regular regimen of manure was enough to meet the needs of soil organisms. But those are just secondary provisions for ensuring that the nutritional needs of the life in the soil are met. The most important aspect is the continuous production of plant roots, which after fulfilling their function while alive die off and accrue as "bacteria food." It's not enough to simply consider the leftover root matter after the harvest, either. You also have to keep in mind that any plant is constantly shedding roots at the same rate as it creates new ones. This means that bacteria food is constantly being produced and finely deposited throughout the soil. The actual quantity of the soil organism nutrients produced by nature cannot yet be determined. What's left of the roots in the soil after the harvest is simply the remainder of the amount that was actually produced.

When a root hair dies it is immediately settled on by microorganisms and consumed (Figure 22). The channel that had been occupied by the root is lined with a coating of humus, allowing it to remain as a water and air channel through the soil.

If insufficient root matter is present, the soil will contain few organisms, meaning that the biological tillage and humus formation will be inadequate and issues with silty or compacted soil will become more pronounced. Causes of this are:

1. The crop rotation is not producing enough root matter. The soil organisms may suffer from insufficient nutrients. Long

Figure 22

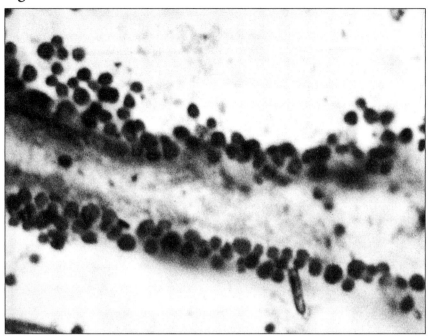

Bacteria breaking down a root hair.

fallow periods always mean sensitive periods without nutrients.
2. In many cases, incorrect tillage in which the densely root-filled upper topsoil mixes with the root-free lower topsoil, causing the already inherently limited supply of bacteria food to diminish each year.
3. An unfavorable lime status or limited nutrient content. The consequences of this are not just a reduced harvest, but also reduced root production.

Crop rotations, especially in intensively worked areas, do not produce enough food for the soil organisms in the form of root remains, and even using manure isn't enough to overcome this shortage. The long fallow periods also mean periods of hunger and vulnerability for the life in the soil. The hunger is due to the lack of root matter being produced and the vulnerability is due to the lack of the necessary vegetation cover to maintain a balanced soil environment, which the soil organisms need in order to thrive. This decimates the life in the soil from year to year and causes a breakdown of the biotic community's interrelationships, a problem characteristic of many fields nowadays that causes a progressive loss of friability and humus. This insight is already enough to conclude what has to be done and how the problem needs to be tackled: the fallow periods must be eliminated because they are contrary to nature and they must be replaced with vegetative cover.

Intercropping is an option in areas where the climatic conditions are appropriate and feed is badly needed for livestock. Since the feed plots effectively shade the soil and are cleared during the lushest period of their life cycle, the result is soil supplied with high-quality nutrients for the organisms living in it. The beneficial preceding-crop effect in a crop rotation is caused by this phenomenon. Unfortunately, intercropping with feed is restricted in many ways by climatic and operational conditions. In dry areas with small livestock holdings in particular, another way needs to be found to bridge the fallow periods. What's needed is a low cover crop that conserves water, covers the soil, and produces as much root mass as possible. This cover crop must fulfill three criteria: cover the ground, create root matter, and introduce variety into the frequently monotonous crop rotation. There are a number of possibilities and different variants for cover crops:

- a. Rapeseed as a cover crop: This method works by sowing the seeds as densely as possible (18 pounds of rapeseed or canola per acre, or 20 kilograms per hectare) to impede the development of the leaves and create a dense root system in the soil. Fertilizing the rapeseed cover with phosphorus and potash is effective because they encourage root growth (particularly the phosphorus). Nitrogen fertilization, on the other hand, is not appropriate here. Nitrogen encourages the plants to grow taller and to form thick taproots. Neither is desirable when working with a cover crop; on the contrary, the goal is to limit the development of the individual plants and instead to create a lush, unified, low carpet of vegetation that permeates the soil with a dense system of unusually fine roots.

As a rule, a rapeseed cover is plowed up in the fall because it frequently becomes too lush in the spring, consuming so much water that it can easily disrupt the emergence of the main crop. But it's entirely possible to leave it over the winter. Due to the cramped space, it usually freezes and can be plowed up in March or early April with a disk harrow. It's advantageous to spread a mist of manure (4.4 tons per acre, or 100 decitons per hectare) over the rapeseed cover before plowing it up, which contributes more (and more varied) nutrients to the soil. This method is especially used before planting root crops.

A rapeseed cover grows so quickly that it's never a problem to find a place for it in the crop rotation. Even the time between planting spring cereals and winter cereals is enough for it to develop as long as it is employed immediately after the harvest. A rapeseed cover is just about the simplest and most effective way to stimulate the soil, but it also requires that certain conditions be fulfilled.

The first and most important requirement: the rapeseed needs at least 6 inches (15 centimeters) of loose soil so that the roots have enough germinating power to mellow the compacted areas. Using shallow skim furrows causes the roots to mostly keep to the upper portion of the topsoil, and as a result the lower topsoil dries out without being mellowed. A cloddy seed furrow and poor growth of the subsequent seed are the results of not plowing deep enough into the topsoil. Experiences with this sort of incorrectly created rapeseed cover lead people to the completely erroneous conclusion that the cover crop has used up all of the water needed by the subsequent crop! A properly employed rapeseed cover creates such mellow soil that the subsequent winter cereal's growth will be assured even in fall seasons with extremely little rain (Lower Austria in 1953 and 1956).

Preparing a rapeseed cover with a stubble plow is certainly a safe method, and you'll have to turn to this machine if a spade sample shows that you have to loosen deep soil. It's been demonstrated that stubble furrows 6–8 inches (15–20 centimeters) deep are sufficient in most cases to produce excellent results, meaning that this method can be even easier. You must be certain to take another spade sample after plowing up the rapeseed cover to determine whether the entire topsoil has actually become friable or whether some areas of compaction still remain. In the latter case you should harvest the rapeseed cover with a stubble plow in order to break open the barrier between the topsoil and the subsoil without burying the friable topsoil through excessive dredging.

The second requirement: the rapeseed cover must not be planted in cloddy furrows because fine-grained seeds require a fine seedbed. A roller should be used if needed.

The third requirement is quite obvious: the leftover stubble must be broken immediately after the harvest. This requires a good deal of tractor power. Keep in mind that a combine harvester delays the harvest by two weeks compared to the usual harvest methods, so the subsequent work must be rushed. But if the stubble breaking is drawn out for this reason and the fields dry out, then the next steps will be completely outside of your control. You simply have to anxiously wait until it rains and eventually make do with an improper

Figure 23

A six-week-old rapeseed cover.

planting, causing you to lose the yields you achieved with the combine harvester the next year.

Figures 23 and 24 show a six-week-old rapeseed cover and its effects on the friability of the soil. You can clearly observe how it thoroughly covers the soil and the dense root system as well as how the increased friability of the soil clods accomplished in a short amount of time.

You can see how the soil loosened by the plow is densely covered by roots with individual root strands already having reached a depth of 23.6 inches (60 centimeters) or more in this short period of time. No other plant has yet been found that can even come close to spreading its roots so powerfully in such a small amount of time.

At this point it's worth discussing the differences between making soil more friable via physical methods and via biological methods. Figure 25 makes these differences clear. Initially, the broken-up pieces of soil are still raw and angular, as shown on the left side of the

Figure 24

Loosening done by the rapeseed as a cover crop

image. The pieces produced by the plow still lack the porosity typical of a crumb layer. The right side shows the same soil six weeks later after the roots have permeated it and broken it up further. At this point the soil is in the form of porous crumbs with a spongelike structure with the biological tillage that is taking place at this point making it friable.

The friability resulting from a rapeseed cover is not inferior to that resulting from a mixed crop that includes legumes. The protein contents of various different catch crops and cover crops were tested, and it was determined that a rapeseed cover works just as well as using red clover in intercropping; both provide high-quality nutrients to the organisms in the soil. A rapeseed cover is also the cheapest catch crop. These benefits are equally true of canola, which grows even more quickly and is a better option for seeds that are sown later than August 20. Mustard, on the other hand, makes a very poor cover crop as it only grows stems and taproots when planted in the density required. This doesn't provide the soil with enough shade or with the network of fine root filaments it needs.

    b. Clover as a cover crop: Sowing a variety of clover that forms tufts along with a cereal crop leaves you with a clover cover after the harvest of the cereal, which is then plowed up in the

Figure 25

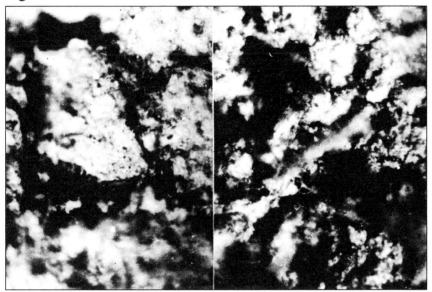

Left: physically broken-up soil. Right: six weeks later, the soil structure created by the rapeseed cover

fall or early spring. Since the clover is already spreading its roots during the growing season, this completely avoids the clod formation that is so common in cereal fields (see also Figure 17).

For lighter soil the best option is white clover (7 pounds per acre or 8 kilograms per hectare), for medium soil the best choice is alsike clover (10.7 pounds per acre or 12 kilograms per hectare), and for hard, lime-heavy soil the best choice is black medick (21 pounds per acre or 24 kilograms per hectare). You should be cautious when using white clover, however, because it is susceptible to clover fatigue. Kidney vetch is not a suitable option because although it does thoroughly cover the soil, it does not make it sufficiently friable. Especially in dry areas you should be sure to only choose varieties that are known to be drought-resistant.

A clover cover also carries significant economic advantages. It avoids the need to break stubble, freeing up manpower and machinery for use in other areas.

A special variant is the clover-straw cover, which is becoming more and more popular in operations that make use of little or no livestock. It makes use of the waste straw from a combine harvester and the finely cut straw from a straw chopper, which is spread over the growing clover. The clover grows through this straw, breaks it up, and builds an ideal interconnection between the living and dead vegetation covers. This combination of living and dead material effectively imitates the natural conditions found in meadows and forests. It's becoming increasingly clear that a dead layer of straw lying on the surface in the shadow of a living vegetation cover provides the most natural possible feeding area for the animals living in the soil. It's downright astonishing how vibrant a community of small creatures can form under a straw layer like this and how significantly they increase the soil's humus production. The creatures in the soil can also very effectively consume the straw in the form of a clover-straw compound as food and convert it into humus. In unshaded soil, on the other hand, the population of small animals is so small that none of the straw is made into humus. Instead, bacteria and fungi quickly consume it fully into carbon dioxide, which is practically useless.

An even more effective option is a combination of high-quality manure with a cover crop. The first step is to finely distribute the manure. Manure spreaders available for purchase now are so good that they allow you to spread a thin mist of manure (4.4 tons per acre or 100 decitons per hectare) as soon as the cover crop begins to spring up from the ground and provide shade. This sort of combination of manure and a cover crop does an excellent job of healing structural issues in the soil and also makes the labor required significantly easier. Even the period of intensive labor that has always been necessary to produce and store the manure is now broken up because the manure can be spread over the course of several months. You can use the manure spreader on the cover crop in the fall and the growing winter cereal in the spring, making extensive use of the manure cover. You don't need to worry about the nitrogen contained in the manure blowing away. Experiments have shown that the ammonium nitrogen is caught by the roof of leaves overhead benefiting the plant.

We've only addressed types of cover crops that we have practical experience with and that demonstrate the goal in mind: forming a vegetation cover that contains a mixture of living and dead material in imitation of nature. There are certainly other options out there, and there are no bounds on the ingenuity of farmers.

A second issue closely associated with improving non-friable soil is tillage. Although two-layer tillage (shallow turning, deep loosening) is best for meeting the biological needs of the soil, it frequently leads to technical difficulties. It's therefore best to try to get by with simple plow types and to choose plow bodies that fit the field. The role of the plow isn't just to loosen the soil and bury the leftovers from the harvest in order to clear the seedbed, but also to place the root matter left over from the previous crop in the location where it can best be utilized as food for the organisms in the soil. You should therefore think of a moldboard as a sort of serving tray. The soil can only be properly fully turned if the entire depth of the topsoil has been thoroughly permeated by roots from the previous crop. In many cases, however, the entire root mass is located in the upper topsoil, with the lower topsoil being compacted and lifeless. In these cases a deep turning buries the food for the organisms and makes the soil more susceptible to silting and compaction the next year. With this sort of soil you need a plow that doesn't fully turn the soil but turns it just enough that the friable upper topsoil remains at the top. In any given field you must use the spade test to decide what kind of plow is appropriate for the local soil conditions in order to reach the correct decision. Equally important is the shape of the plowshare, because it's important not to separate the topsoil from the subsoil with horizontal cuts. Instead, you want to produce furrows that are as rough as possible at the bottom.

We'll address the roles of a balanced lime state and of providing the soil with a regular supply of nutrients in maintaining friability later; for now, we'll just say that lime improves the soil's structure through its coagulating effect and that a sufficient supply of nutrients allows a dense root network to grow in the soil, which provides the soil organisms with ample raw materials for biological tillage and humus formation.

Non-friable soil certainly can't be healed overnight. Particularly difficult "patients" can't even be healed over the course of a full growing cycle. It's not uncommon for weather conditions to be uncooperative, limiting you to partial success. In these cases you'll find varyingly friable large clods in the topsoil despite all your efforts, and some remnants of compacted areas will remain present. It's common for the soil to partially lose its friability during the next growing season, with signs of structural breakdown reappearing. But you can always count on an improvement in the soil conditions and a corresponding increase in yields, which already makes your efforts a success. If you continue to apply the healing measures, however, you will be able to fully repair compacted soil in time. You just need to make a point of first thoroughly examining the soil via the spade test and making a brief record of the cross-section of the topsoil. Then you should consider how to approach the healing process.

As a rule, the soil in any plot can be healed to a greater or lesser extent, so it's important to design a healing plan that takes on a por-

Figure 26

Cross-section of the topsoil before (right) and after (left) its lack of friability was ameliorated.

tion of the field each year, remedying all of the plots over the course of the rotation.

Figure 26 shows a field that had been exhibiting significant soil compaction and was healed with the use of a rapeseed cover. It increased the wheat yield from 1.1 ton per acre to 1.8 tons per acre (25 decitons per hectare to 40 decitones per hectare).

In addition to the loss of friability characterized by the compaction of the topsoil after it is plowed—often incorrectly referred to as hardpan—actual hardpan is also an issue. It was an especially common issue in the era when plows were led by teams of animals because these plows reached the same depths every year causing the bottoms of the tracks of the plowshares to form a compacted layer often several inches thick. This layer acted as a sharp border between the topsoil and the subsoil. This is another example of a barrier layer keeping the roots from reaching the subsoil, even though its cause is different. These layers must be physically broken up as part of the healing process and then quickly provided with the right conditions to create friable soil in order to give the dead soil new life.

Chapter 5

THE SOIL AS A WATER RESERVOIR

The end goal of maintaining soil friability is to ensure a suitable water balance between the soil and the plants. Water and air channels can only maintain their functions over a sustained period in healthy, friable topsoil. The most important thing of all, however, is that it remains possible for water to move between the topsoil and the water reservoirs in the subsoil so that the field can quickly absorb any rainfall and successfully store it in the subsoil. This is crucial because in dry times a field that has stored the most moisture in an accessible manner has the best chance of survival.

How does healthy soil function as a water reservoir? How can the water balance be disrupted in a non-friable field?

First of all we need to correct an old belief. Plants are not supplied with groundwater that rises via capillary action if their roots don't directly grow into the groundwater area. Capillary water transport does indeed occur, but it's much too sluggish to meet the plant's water needs. It is therefore important to differentiate between groundwater soils and rain soils and to note that in the case of the latter the plant can only use the water stored in the area reached by the roots. The amount of water stored is thus signified by how deeply the roots reach. It's a little known fact that annual crops reach a depth of nearly 6.5 feet (2 meters) under optimal conditions, allowing them to make use of water stored down to that depth. Even cereal crops, which form a flat, clump-like mass of roots and are accordingly known as "flat rooters," reach these depths with a few lightly branched root filaments. The left side of Figure 27 shows

Figure 27

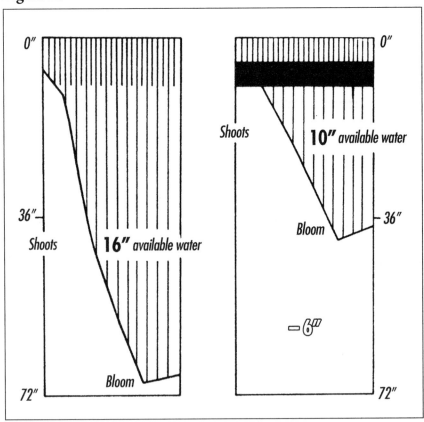

Root development and usable rainwater storage in healthy and sick fields.

the optimal progression of a cereal root. As soon as the plant's longitudinal growth sets in, the root starts to move downward at a high speed (0.8–1.1 inches or 2–3 centimeters per day), and by the time it blooms it will already have reached its optimal depth. However, this is only possible in healthy, deep soil. As soon as the soil structure breaks down due to a lack of tillage, the root will inevitably be blocked from growing any deeper, as the depiction on the right side shows.

It's not the physical resistance that stops a root from growing in compacted topsoil, but rather the lack of air that comes with it. Quite understandably then, you will never achieve optimal root depth in a sick field. In the case at hand the roots only reached 3.6 feet (1.1 meters), meaning that the supply of water available to the plant is considerably shallower than it could be. In any given case you can determine the water capacity of the reservoir area to figure out how many inches of rainfall will be stored in the root area in a usable form. This important figure is referred to as a field's rain capacity. In the example shown the healthy field has a rain capacity of 16 inches (400 millimeters) while the sick field has a capacity of just 10 inches (250 millimeters). A field may enter the growing season with sufficient moisture from the winter, but its level of health will determine whether it has 16 inches or just 10 inches (400 millimeters or just 250 millimeters) of water available to it in the summer. Clearly, healthy soil is better able to withstand periods of drought and will provide bigger harvests than sick fields. In this case, a loss of friability means sacrificing 6 inches (150 millimeters) of rain!

In this case, we tested deep, loamy soil which inherently functions as a large water reservoir for the plant. But the amount of usable water stored can differ significantly from field to field depending on the structure of the subsoil. While a rain capacity of 16 inches (400 millimeters) represents an optimal value, it can be as low as 4 or 2 inches (100 or 50 millimeters), or even less in shallow soil. In some cases, you can simply excavate vertically to determine how deep the roots will penetrate into the subsoil. For each 4 inches of a given layer, the following amounts of rain are stored in a usable form:

Coarse sand	less than .2 inches	of rain
Fine sand	.2–.4 inches	of rain
Loamy sand	.4–.6 inches	of rain
Sandy loam	.6–.8 inches	of rain
Loam	.6–.8 inches	of rain
Loamy clay	.4–.6 inches	of rain
Clay	less than .4 inches	of rain

You can use these values to determine the rain capacity of any field; they are a reliable indicator of how much a particular soil type can hold. The larger the useable water reservoir in the subsoil, the less the plant will depend on the randomness of summer rainfalls and the more viable measures taken to aid in its cultivation will be. **It is not the surface area, but the volume of the root-filled area of the soil that matters!**

The question now arises: how much winter moisture could a field store in theory, and how much does it store in practice? How deeply roots can penetrate into the subsoil and how deeply they actually do has been determined from many individual cases. Comparing the possible and actual rain capacities can tell us how much more reservoir space in the subsoil can be exploited if the field is healed (i.e., if the connection between the topsoil and the subsoil is reestablished).

The following findings have been statistically evaluated to determine that, due to compacting,

in 11 percent of cases	more than 6 inches (150 millimeters) of stored rainwater,
in 24 percent of cases	4–6 inches (100–150 millimeters) of stored rainwater,
in 37 percent of cases	2–4 inches (50–100 millimeters) of stored rainwater,
in 28 percent of cases	less than 2 inches (50 millimeters) of stored rainwater

remain unused because topsoil compaction has impeded the roots from penetrating downward. Forgoing this winter moisture is particularly irresponsible in dry regions, but it always leads to uncertainty even in areas that get plenty of rain.

Topsoil compaction doesn't just impede the exploitation of underground water sources, it also impedes their replenishment. It's important for the soil to be able to store fall and winter rainfalls without any losses. It's common, however, to see water backing up in the upper topsoil because a barrier layer is delaying it from sinking into the lower topsoil. Instead of being stored in the subsoil for the next growing season, most of this water is squandered via evaporation or lost when it flows away on the surface. A rain-heavy autumn may spark hope that there will be sufficient moisture in the winter. People console themselves with the thought that even if their fields remain impassable to machinery for weeks at a time so that fall crops have to be skipped over they can at least expect to have enough moisture for a spring planting. But if they were to dig up the soil, they'd be disappointed to learn that only the upper topsoil is wet and dry subsoil lies underneath compacted topsoil. The

Figure 28

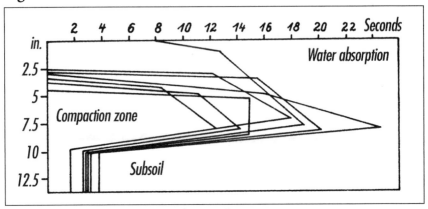

Water congestion in compacted topsoil

supposed winter moisture isn't there at all! This sort of situation is characteristic of non-friable fields and paints a very telling picture of their circulation issues. We've already pointed out that compacted areas of the topsoil are less porous and thus less permeable than the subsoil. Figure 28 shows water blockages in the compacted topsoil of various plots in order to illustrate how much more sluggishly compacted areas of the topsoil absorb and distribute water than the subsoil.

It's very easy to obtain an instructive example of these circulation issues by taking a spade sample to check the water absorption of the topsoil and subsoil. You just need to drip a few drops of water onto the same spot and count how long it takes before they are fully absorbed. This very simple experiment helps to get a better idea of the degree of soil compaction and usually reveals that the subsoil could absorb the rain five or even ten times faster if there were no barrier layer in the lower topsoil. Even skeptical farmers who don't think much of the concept of friability have realized that something isn't right. Only healthy soil can quickly absorb rainwater and distribute it so that losses due to evaporation are kept to a minimum.

These findings present a very different outlook on the dangers of drought. We can't avoid that sometimes there will be either too much or too little rain, hampering or delaying important work on the field. Proper soil care can largely alleviate these difficulties, however, by ensuring that your soil can fulfill its function as a balancing water reservoir. Drought impedes field cultivation just like soil compaction. It's simply a consequence of the breakdown of the soil's earlier strength, and making the soil friable again is an effective weapon in the fight against drought. It is especially important to realize this if you are planning an irrigation system to help with these issues. You should closely inspect your fields beforehand and see what the rain storage situation is. Generally, you'll find that the natural rainwater reservoirs in the subsoil aren't working the way you expected, and that you can exploit and make usable a substantial store of water for free by alleviating its structural issues. Not until then, when the natural water supply is being fully exploited, is it sensible to risk the expense of an irrigation system, which should primarily be employed in situations where the natural water stores don't catch enough of the winter moisture because the soil is too shallow.

The lower the rain capacity of the field, the more return an irrigation system will provide on your investment. But this is only true if you've already eliminated all circulation issues in the topsoil and it is stable enough to withstand the heightened pressure of water from the nozzles. If this barrage of water batters the field and becomes blocked up in the upper topsoil, then you haven't accomplished much, and may even have done more harm than good. An artificially watered field therefore needs its own special regimen of care or else most of the introduced water will just evaporate again. Furthermore, you must consider the following: a largely dried-up field first requires a certain level of moisture penetration before the plants can extract and use water from it. Any given soil has a certain "critical water content" under which it can no longer deliver water to the plants. The pores close up and can survive long dry periods to an extent that depends on their "drought resistance." The longer you wait to apply the water, the more water will be necessary to get the soil back to its critical water content. The whole effort can therefore be a losing proposition, and it's been shown in practice that it's a good idea to establish an irrigation system whose cycle is interrupted only when natural rainwater appears. This both provides a balanced supply of water to the plants and mitigates the silting effect of sprayed water. Always remember that the soil will get siltier and siltier the wetter it becomes.

The water content of friable soil is characterized by its balance. In compacted soil, on the other hand, it fluctuates from one extreme to the other, alternating between too wet and too dry. This inconsistency poses two dangers to the plants. Plants primarily use their stomata to regulate their water consumption. But long periods of drought lead to a paralysis of the stomata, robbing the plants of their natural ability to regulate. This causes permanent damage, and water that is applied too late is no longer able to help much. The other danger is that any inconsistency in a plant's water supply weakens its constitution which takes away from its natural resistance to pests and disease. It's repeatedly been observed, for example, that fusariosis in cereal crops from compacted fields causes much heavier losses than in healthy soil. The infection may be the same in both cases; the difference lies in the fact that it is limited to the surface of the stalks in healthy fields whereas the fungal hyphae infiltrate the vascular systems of plants in unhealthy fields, preventing crop development. Even heart rot and dry rot seem to be alleviated by a balanced water supply. Animal pests such as weevils, flea beetles, pollen beetles, etc., cause significantly greater damage to unhealthy soil than to healthy soil. Taking care to maintain friability improves plants' natural resistances which makes it easier to employ any measure to protect them. The fact that the fight against disease and pests constantly requires more and more effort is certainly partially due to decreases in natural fertility. Plant care and soil care thus have many points of contact.

Soil and plants don't just need the rainfall that refills the dry water reservoirs from time to time, however—both also require the moisture provided by dew on a daily basis. People have long undervalued the importance of dew formation, perhaps because the quantities supplied are much less than in rainfall. But the amount is not so important—much more important is that its formation via condensation of water vapor takes place in the soil and on the surfaces of plants instead of high up in the clouds. This produces warmth in the immediate environment of the plants which reduces the danger from frost. It also works against the surface of the soil drying out and triggers a daily stimulation of the soil organisms in the exact location where there's a risk of crust formation. It's crucial however that the green plant parts keep their stomata open for as long as they are covered in dew allowing them to absorb it without transpiration.

Dew formation certainly depends on the effective surface area of the dew receiver: crumb structure. Friable soil absorbs dew more effectively than non-friable soil. All farmers will be able to confirm this from their own experience. It has not been proven but it's very likely that dew can keep the entire topsoil under a crumb cover moist, even if no rain falls at all. The "organic crumb cover" that develops through soil care seems not only to be a highly effective protection against evaporation, but also an equally effective dew collector. It may be interesting to note in this context that a low rapeseed cover provides the soil with (.03–.05 inches) of condensation in the form of dew per night.

Chapter 6

THE SPADE TEST

Every farmer needs to be able to make an informed decision about what sort of care regimen the soil needs. For this reason, the existence of a method for field diagnosis that allows you to ascertain the soil's needs with the use of only the very simplest aids represents an enormous step forward. J. Görbing developed the spade test over the course of decades of experience. The test makes it possible to uncover and check the condition of the soil at any time, and it's so simple that any farmer can carry it out. The expression, "The farmer is his soil's personal doctor" is not an exaggeration. A soil specialist should be consulted in complex cases, but determining how deep to plow, what plow body to use, and many other details of how to work a field—as well as checking the state of the life in the soil—can all be accomplished by the farmer using only this method.

All you need is a solidly built flat spade with a rectangular head (⅞ inch by 12 inch, or 18 by 30 centimeters) and a strong T-handle. You also need a three- or four-pronged claw and a yardstick. It would be a good idea for plow manufacturers to also produce these aids and package them together because spades and plows belong together on the field.

The goal of the spade test is to get a look at a cross-section of the topsoil that is as undisturbed as possible by excavating a brick of soil, as portrayed in Figure 29.

Insert the spade vertically into the soil with side-to-side lateral movements (1), avoiding any forward or backward pressure on the soil in the process. Then, with the help of a second spade, excavate a hole in a way that uncovers the first spade (2) and then stick it back into the ground (3) so that a brick-shaped cross-section of the topsoil is obtained after tipping it over. Breaking the brick of earth free is much easier if you cut two vertical grooves (2, 3) into the wall created after digging the hole. Now, closely examine the brick of soil you've dug out by placing a yardstick on the brick and dissecting it with the claw (4) from the top down. First clear away the crumbly, friable soil and look for embedded clods. Compacted areas will remain in the interior of the topsoil, and you should carefully break these up by hand so that you can determine the degree

Figure 29

The Görbing spade test.

of the compaction and differentiate between clod-like and plate-like breakage. As you dissect the brick of soil the plant roots will become uncovered on their own, making it easy to observe how the roots react to different types of compacted soil and how it impedes their development. With this in mind you should take the sample from somewhere a plant (a crop or a weed) will be encompassed within the soil brick. In each case, there will always be a close relationship between the condition of the soil and the form taken by the roots:

1. Friable soil is filled with a rich root network; individual root filaments are stretched out.
2. Less friable soil (compaction level 1) doesn't contain as many roots; the individual root filaments are not stretched out, but finely curled.
3. Soil that breaks into clods (compaction level 2) only contains a few roots; most of the roots are located on the surfaces of the clods (spade roots) or in wormholes. The individual roots are more widely curled depending on how big the clods are, which the roots must have surrounded
4. Soil that breaks into plates (compaction level 3) has no or very few roots in the vertical direction. Roots can only be found along the horizontal edges of the clumps (in crevices).

Figure 30 gives the cross-sections of three topsoil samples along with their root systems as examples. You should record the cross-sections of individual fields annually in this simple way. These records will show you the life progression of a field over the course of multiple years, and you can use them to follow the appearance and disappearance of symptoms of poor health and get to know the soil better and better by seeing the effects of various cultivation strategies. The more closely you observe and compare the good and bad

Figure 30

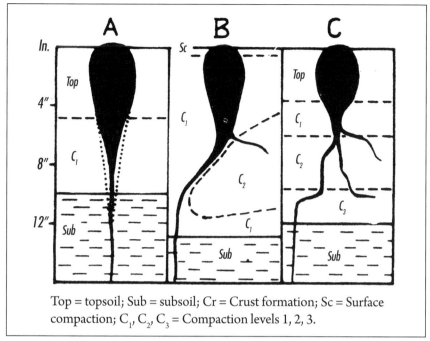

Top = topsoil; Sub = subsoil; Cr = Crust formation; Sc = Surface compaction; C_1, C_2, C_3 = Compaction levels 1, 2, 3.

Schematic depiction of the structural findings in a beet plot with the Görbing spade test

areas of a field to each other, the more you can learn from them and your experiences.

Section A shows the cross-section of nearly healthy soil; it's friable (top) up to 5 inches (12 centimeters), less friable (C_1) up to 10 inches (25 centimeters), and underneath that is the unworked subsoil (dashed in the drawing). Compaction level 1, characterized by crumbly breakage, is already enough to impede secondary growth somewhat. The way it should look ideally is shown with dots in the drawing. Even this limited degree of compaction is enough to diminish your yields.

Section B shows a field that has been too plowed with furrows that are too wide. It also exhibits surface compaction (Sc). Most of the topsoil is friable (C_1), but a chunk of soil left by the plow that breaks into clods remains embedded within it and exhibits the second degree of compaction. As you would expect, you can perceive a clear division within the beet root. The main root is forced to work around the compacted area, and a lateral root becomes anchored in it. This type of malformation is already enough to significantly reduce yields.

Section C shows a very difficult case. Here, compacted areas have built up above each other like the stories of a building (C_1, C_2, C_3); they grew upward from the bottoms of the furrows. You can see how the beet roots react to this obstruction. Level 1 compaction begins at 4 inches (10 centimeters), and the beet root tapers off. Level 2 compaction begins at 6 inches (15 centimeters), and the root branches out to try to find a way through the cracks and root holes. Note the coarse, wavy form of these roots. The detour they were forced to make can clearly be recognized from this shape. At 9.4 inches (24 centimeters), the root hits a level 3 compaction and has to branch out horizontally, trying to happen upon areas where it can continue deeper here and there. There's no need to go into the specifics of the yield reduction caused by this sort of root situation; it means a bad harvest.

To be able to judge the structural conditions of a field you have to be sure to carry out multiple spade tests at a time. As a rule you should see very consistent results, but there are also cases of significant irregularity even within the smallest areas. These irregularities are often the result of sparsely planted plots. Carelessness during the planting can also cause an uneven structure. In any case, you have to keep in mind that the spade test only reveals a snapshot of the progression of structural changes in the topsoil. It's not until you've performed multiple tests over time and compared the results that you have a useful picture of the progression of life in the soil and can start to really get to know it.

It's rare to find a farmer who reacts indifferently to a spade test. Many very experienced farmers have claimed that the spade test was a major step forward for them in understanding their soil. You simply have to look into the soil and concern yourself with the processes taking place within it.

Chapter 7

ORGANIZED HUMUS MANAGEMENT

Humus management refers to feeding soil organisms in a planned manner.

This definition assigns a substantially broader role to the term than has been used before. Humus management is no longer limited to manufacturing and applying manure, but also now involves creating favorable living conditions for microorganisms. The core issue in humus management lies in the coexistence between plants and soil organisms ordained by nature.

There are three tasks related to humus management in a field:
1. Feeding soil organisms organic substances that they are able to break down (friable humus).
2. Stabilizing the soil structure with use of a water-resistant binding agent (stable humus).
3. Productively exploiting the organic substances available (carbon turnover).

Figure 31

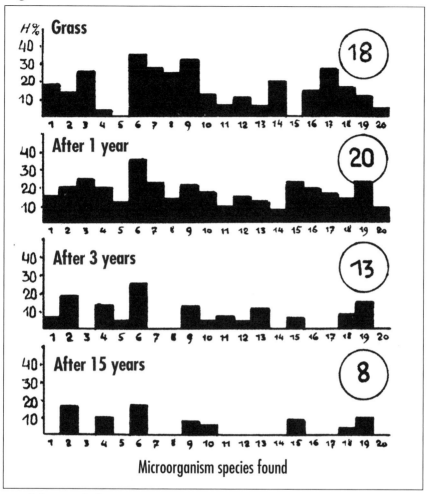

Breakdown of a biotic community; number of microorganism species found in substantial numbers.

The main goal is to consider the nutritional needs of the soil microbes and small animals and to make enough organic material available to them. Any measures taken in the areas of crop rotation, fertilization and tillage must have increased root production as their aim, as this is the way to solve the friable humus problem. The more numerous and diverse the available nutrients are, the more effec-

Organized Humus Management

tively the microbes will be able to multiply. This is extremely important, as you can see by comparing the breakdown of the biotic community in the soil between a field and unworked grassland (Figure 31), which shows how lightly populated soil that has been under cultivation for fifteen years is.

In these experiments the individual species of microorganisms were identified by numbers instead of by their names. Initially, eighteen different types were present in large numbers, which temporarily rose to twenty in the first year of cultivation, but then rapidly dropped to the point where only eight were weakly represented by the end of the series of experiments. However, a planned feeding regimen can cause the number of microbes to increase. Figure 32 shows the composition of organisms present when various different methods of caring for the soil are employed.

When breaking stubble after a wheat crop only four different microbes were found, and in limited quantities. This number rose to eight after manure was added, while a rapeseed cover produced seven different types. Using manure followed by a rapeseed cover increased the number to thirteen. The number of types grew significantly when a light manure mist was spread over the rapeseed cover, with a total of twenty-two well-developed species found.

What is categorized as stable humus? How do these dark-colored colloidal substances form, which, like clay minerals, have the ability to bind plant nutrients and to produce, together with the clay minerals, a clay-humus complex, an especially water-resistant binding substance in the soil? The incorporation of these humus bodies into the soil structure makes its mortar substance more flexible so that when water hits the soil larger pieces will form, which won't be as heavily affected by its silting effects as the small aggregates of pure mineral soil.

Figure 32

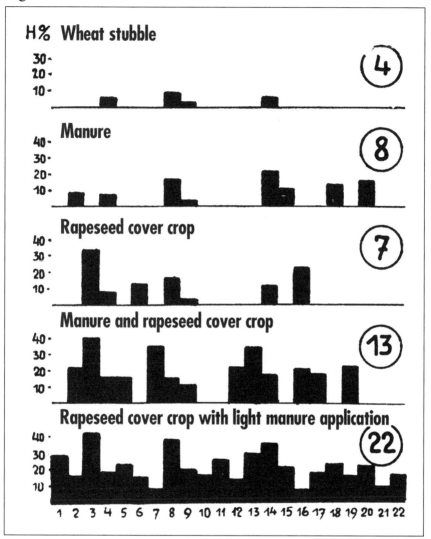

Buildup of a biotic community

Until recently, stable humus was commonly believed to be a residual product of the organic substances that had accrued in the soil degrading, meaning that the humus content of the soil could be maintained or augmented by adding sufficient organic material. People tried to create a substitute for the organic matter removed during the harvest and developed the now-common techniques of using manure and green manure. More and more the humus issue became a manure issue. Certainly friable humus and stable humus formed on the piles of manure, but that wasn't yet enough to be sure that both forms could be usefully exploited in the soil. The quantities produced were also too low to permanently meet the demands of the soil. The soil itself only produces humus when the necessary conditions are fulfilled.

Up until now we've only addressed managing the lives of plants, not their deaths! But it's exactly this stage that is crucial for the formation of humus. How does nature react to the presence of dead plants in the soil? Dead plants rot as they build up, decaying by means of the actions of the soil organisms, a category that includes not just the various microorganisms present (bacteria, fungi, actinomycetes, etc.) but also an equally wide variety of lower animals in the soil (worms, mites, springtails, larvae, etc.). These microflora and microfauna form a tight biotic community in which the constituent members mutually help to break down and consume the dead plants. The decomposition process is depicted in Figure 33.

Two major discoveries related to humus formation have been made recently. H. Franz, W. Kubiena and W. Laatsch succeeded in proving that precursors to humus form in the intestinal tracts of the animals in the soil. These creatures also consume fine soil when they're looking for food allowing stable humus to form while still inside their intestinal tracts, which is later finely distributed through the soil in their excrement.

Another group of researchers, F. Scheffer, W. Laatsch and W. Flaig, proved in their experiments that plant microorganisms are also capable of synthesizing true humus bodies. This is a very essential discovery because, though the role of microbes in biological tillage was already known, no one had realized that true humus mate-

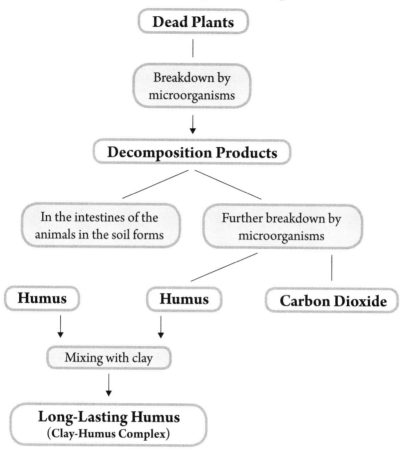

Figure 33
Breakdown, conversion, and composition of organic substances

rial was involved in it. It still remained to be discovered under what conditions the local microorganisms could process the nutrients given to them into humus. F. Sekera was able to show on colored photomicrographs that microorganisms do cause humus to form, and he determined that plant remains are processed into humus if animal material is also available at the same time. This proved that exogenous humus formation via the plant microorganisms in the soil takes place alongside the endogenous humus formation carried out by the animals, and that the formation of new humus in the soil is closely intertwined with functions of the biotic community of microflora and microfauna present.

But how does the humus formation by microbes actually occur? To find out researchers provided soil with various different nutrients. Some areas received only nutrients from plants (roots, green plant mass, hay), some only nutrients of animal origin (insect larvae, mealworms, earthworms), and some areas of soil received both mixed together. Coloration would prove that humus had formed. Microbes supplied with only one form of nutrition or the other were never able to form any humus. Even in the last stages of the breakdown of the nutrients, no trace of any coloration could be found; the food was completely consumed into carbon dioxide. This was not the case when both nutrients were supplied. The plant tissue displayed a localized brown coloration after five days in most cases, and after ten days in every case, which proved to be new humus forming.

In soil where only green rapeseed matter was introduced everything except for the leaf veins had been broken down within twenty days. In soil where dead worms were introduced alongside the rapeseed material, however, partial brown coloration set in after ten days and was fully brown after an additional ten days.

It's not yet known why the organisms require additional animal-based nutrients to produce humus material. It's likely that the corresponding nitrogen supply provided by animal protein enables humus formation.

The density of the life in the soil increases when provided with mixed nutrients in a manner analogous to the humus formation. To test this an experiment was carried out with black medic (*Medicago lupulina*) and mealworm larvae (*Tenebrio molitor*); the results can be found in Figure 35. When only plant nutrients were provided,

Figure 35

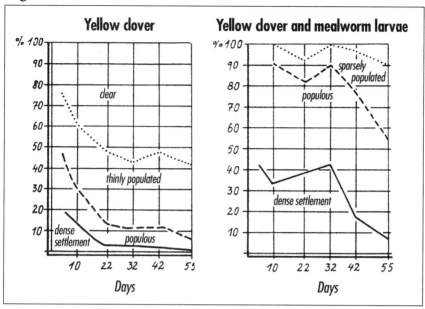

Percent distribution of soil organism settlement density.

the abundance of dense settlement had already sunk to 14 percent after twenty-two days, then remained at a minimum value for a long time. With mixed nutrients, on the other hand, 54 percent of the observed areas were still densely settled after fifty-five days. The up-and-down oscillation of the settlement curve is not due to experimental errors but was in fact always observed with mixed nutrients. This phenomenon is apparently the result of an antagonistic effect. Additionally the community of organisms was much more diverse when given mixed nutrients rather than only animal or only plant nutrients.

Providing the organisms in the soil with a diverse range of nutrients has another invaluable advantage. Contrary to all expectations, the substances decomposed substantially more slowly when a large quantity of organisms were present than when only one kind of nutrient was provided. The by-products and humus precursors that accrue over the course of new humus formation apparently regulate the soil organisms' carbon turnover by slowing down the unproductive consumption of organic substances, thereby allowing the organisms to last longer on the nutrients available to them. The humus bodies are thus the active substances ensuring productive carbon turnover in the soil, while the leftovers from the harvest are quickly consumed if not enough humus is formed, causing the activity of the organisms to sharply drop off after a temporary increase. This "biological straw fire" can therefore be slowed down with improved humus formation and the accompanying productive exploitation of nutrients by the organisms.

But even the most abundant supply of nutrients will do nothing for the organisms in the soil if they are not provided in the right location. **Where do soil organisms prefer to feed?**

The answer to this question as far as earthworms go was attained through the following experiment: soil from a field was put into boxes. Half of each box contained organic matter evenly distributed throughout the soil, and the other half contained the same amount of organic matter but spread out on top of the soil and only mixed in very shallowly so that a crumb cover richly intermixed with organic matter formed. After four weeks, and again after eight weeks, ten earthworms taken from the same field were introduced into each box. Distributing the organic matter in this manner gave the worms a choice between taking food from the interior of the soil or from the surface.

The behavior of the worms was unambiguous. If they were given root matter as food, they didn't care whether it was in the interior of the soil or spread over the surface. In the cases of leaf matter and manure, they preferred it from the surface. They visibly avoided green matter.

These experiments confirmed the findings of H. Franz who dug back out and examined manure nesters that had been introduced into a field and observed how small animals had settled. Contrary to all expectations, microorganisms from the manure pile kept living in the soil, while the indigenous ones didn't accept the food given to them. Along similar lines, K. Maiwald observed that microorganisms from the manure could keep living in the extracted manure while almost none of the local microbes that were added survived. Organic matter introduced deep into the soil, regardless of whether it's manure, green matter, or remnants of roots or stubble, remains in the soil for a long time as a foreign body without serving as food for the soil organisms, instead simply consuming itself without providing any benefit.

Taking the correct biological measures for your field isn't too difficult as long as you take nature as a model. In natural grassland dead matter is present alongside living matter, and as a new plant develops a dead one will also be breaking down. This process forms a layer of plant litter (or duff) under which you can observe all of the stages of decomposition up to and including its final transition into soil. There is a vibrant living world in the soil underneath, a coordinated community of microflora and microfauna. It's worth noting once again how the creatures in the soil, especially the earthworms, loosen and mix up the soil. The same thing that takes so much effort to accomplish via labor—creating the crumb structure—is carried out naturally by the animals in the soil. And microbes contribute to these crumbs and coat them with layers of humus, making the soil in better tilth than can be accomplished otherwise. These tiny workers can also distribute the humus more finely than any machine could.

In a field where these natural conditions do not prevail and are becoming more and more distant due to the effects of monoculture, the periodic drying of the topsoil, poor water circulation, an insufficient supply of root matter, constant disturbances to the topsoil due to plowing, and various other hindrances, the top priority in its cultivation must be finding a way back to a state of nature. This can only be accomplished if the reduced population of soil organisms in the field is reestablished by mobilizing the creatures and microbes in the soil as humus producers. The decisive factor is providing the "microflora-microfauna community" with food in an appropriate location so that they can abundantly multiply and do the work we want them to.

The small creatures living in a field can also be stimulated by not simply plowing the surface plant remains and manure under, but instead shallowly introducing them into the soil so that a crumb cover rich in organic matter forms. This crumb cover replicates the natural layer of plant litter that exists in meadows and forests. It provides the flora and fauna in the soil with protection, a living space and a place to feed.

This organic crumb layer seems like it will be the central core of future humus management because:
1. If you manage to get especially intensive biological tillage going in the crumb structure where the rain hits and causes most of its silting effects, it makes all measures taken to keep the soil porous easier.
2. If you can densely populate the field with animals and microbes, they will
    a. produce stable humus,
    b. regulate the carbon turnover, and
    c. continuously help loosen the soil.

The crumb layer needs to become a permanent part of the field's topsoil, just as layers of plant litter form a component of the soil in nature, and it can be provided anytime organic matter is added to the soil. Always remember that root matter belongs in the soil and leaf matter belongs in the crumb layer.

Stubble furrows provide an opportunity to create an organic crumb layer because they only blend the stubble remains very shallowly into the soil.

A cover crop, however, is an especially effective tool to improving your soil. Creating high-quality root mass to help make the soil more friable is only a part of its function. Its second but no less important benefit is in turning leaf material into an organic crumb cover that will survive in the field until the next crop is planted. This is the best way to guarantee that humus will form, and it will ensure a porous surface for the next crop. If the next crop is a root crop, using a cover crop will make the otherwise laborious hoeing significantly easier and will provide better conditions for mechanical separation work. Preparing an organic crumb cover for a root crop plot means hiring an army of hard-working microscopic "farmhands," as no other equally cleanly implemented hoeing can loosen the topsoil as well as the animals in the soil.

Keep the cover crop fairly low to keep too much leaf material from growing. In cases of frequent rainfall or especially fast-growing plants, however, the leaf cover frequently grows too high. In these cases, you can help the situation by driving a cutter bar over it and leaving the green material on the field. This "mulch layer" is another step toward combining living and dead matter. Alternately you can drive over the field with a fast disc harrow before tilling, breaking up the cover crop and causing some of it to wither, which reduces the bulk of the leaf material. You must still be careful to shallowly introduce the green matter. Plant parts still visible here and there on the field are far preferable to forcing everything into the soil with a plow, leaving foul-smelling bunches of green material still present the next year.

And how should manure be applied? Until recently, if you spread the manure by hand, even very meticulously, it was impossible to keep fist-sized lumps of manure, which are useless to the life in the soil, from turning into peat or mineralizing. In many cases, manure exploitation resembled manure destruction. No other equally labor-intensive strategy could change the fact that you could only expect a partial effect due to shortcomings in the manure's application and placement.

The manure spreader has solved the problem of evenly distributing the manure. It significantly raises the productivity of the manure. The way the manure spreader works provides many advantages, the most significant of which being that the concentrated period of heavy workload is now spread out over a much longer timespan. As long as the manure was being spread with a pitchfork, tillage had to take place immediately. This is no longer the case, and now the resulting fine blanket of manure provides protection to the organisms in the soil. This coupling of manure application and placement also avoids the risk of the manure becoming greasy under poor soil conditions. From the standpoint of soil care, spreading the manure as a "manure blanket" over the plants represents an enormous step forward. It provides an excellent imitation of the formation of a natural layer of plant litter.

Preparing the manure with chopped-up plant litter represents an additional improvement. It both aids in mechanizing the manure process and increases the quality of the manure by absorbing as much urine as possible while inside the barn. Additionally this advance has opened up further possibilities for significantly simplifying manure management. Many have asked whether manure that retains almost all of the manure slurry (i.e., has a narrow C:N ratio) needs to mature any further on the manure pile. To put it differently: can nitrogen-rich chaff dung be applied to a field without rotting?? This question is of great practical significance because stockpiling manure in piles always causes organizational issues and is associated with unavoidable losses. Rotting manure is essentially a waste of friable humus, and it would make things much easier if you could load manure directly from the stable into the manure spreader, or at least cart it out into storage clamps. But how does the soil react to the fresh manure drawing on its nitrogen? It's well known that unrotted manure temporarily draws nitrogen from the soil, rendering it unavailable to the plants and thus diminishing yields. Don't assume that the higher urine content of chaff dung will offset this detrimental effect. The same diminished yields are seen with manure from mountainous areas, which are very low in plant litter, even though the fresh manure has a better C:N ratio in this case. But this seems to be an issue only if you apply the manure deeply into areas where it will compete with plant roots. If you mix it in with the crumb layer, however, the manure will finish rotting on the surface layer of the field and will thus barely influence the downward expansion of the roots. Note, however, that experiments in this area have not yet been completed and we do not recommend applying immature chaff dung until this is more thoroughly investigated.

The transition from straw manure to chaff dung can't be accomplished overnight. It frequently depends on the farm's structural conditions. Until you can reach that goal, however, you should work on an interim solution. Cut straw is already enough to bring significant benefits both in barns and in your manure stores. It particularly helps with applying the manure to the soil, which has to be done as shallowly as possible. Manure should not be any deeper than 6 inches (15 centimeters) under any circumstances, so you have to use a stubble plow. Straw manure always leads to difficulties when it's applied, and it runs the risk of just lying in the soil in bundles which causes it to not just uselessly consume material as described above but also to inhibit water circulation and root growth.

High-quality manure can also be created in deep litter. The livestock stepping on the straw breaks it up and works in the urine and feces evenly. The rotting accelerates, producing a narrow C:N ratio.

With this knowledge it seems likely that future humus management techniques will be heavily influenced by the organic crumb cover. Developments thus far have brought us very far from a natural humus balance, and now we must move back toward nature. It should be mentioned that an organic crumb cover is not a completely new concept; it had precursors in other countries. Edward H. Faulkner was already advocating a similar surface treatment in grain-farming regions of North America in 1943, and A. Manninger was doing so even earlier in Hungary. Even W. Laatsch, who was responsible for major advances in humus research, saw shallowly mixing organic substances into the soil as an advance in its ability to form humus. Even though these foreign discoveries can only be applied cautiously in our region, the fact that multiple researchers have independently reached the same conclusions does provide a certain confidence.

Caring for a field's surface naturally is undoubtedly a decisive factor in the soil's fertility, but it's no less important to provide the same level of care to the topsoil and subsoil. Care for the surface and care for the deeper soil must therefore be correspondingly intertwined: **humus care and tillage thus share many of the same tasks**.

We've explained that humus materials need to form on their own in the soil for them to strengthen the soil's structure. Attempts at producing soil-ready humus were not successful. A mineral clay substance (loam) was added to the manure pile achieving the formation of clay-humus complexes. Composting works in a similar manner. But this finished stable humus was not incorporated into the soil structure. Researchers observed the pieces of compost and manure that had been introduced into the soil for two years, ultimately finding that microorganisms (primarily Actinomyces) were indeed densely settling the areas around the pieces of humus and breaking down the organic substances, but no incorporation of the humus into the soil structure was noted.

Solely emphasizing improvements to soil structure would make for a one-sided treatment of the humus issue; plant nutrition must be taken into account as well. As W. Laatsch shows, humus material contains nitrogen, phosphoric acid and important trace elements in organically bound forms, and it provides these nutrients to plants in biologically determined dosages as the humus is replaced in the soil. This decomposition takes place more quickly on warm days than on cool ones, meaning that the delivery of nutrients to the plants always adjusts itself to fit its requirements for the day. There's never an excess supply, and you never get the "nutrient piles" that are almost unavoidable when fertilizing. This biological balance between nutrient demand and nutrient supply not only affects the health of the plant, but is also very significant for its future nutritional value. Humus is the ideal carrier of nutrients.

Checking the soil's conditions for fertilizer has already become common today; it's impossible to apply fertilizer in a coordinated manner without doing so. Unfortunately there has been no similar foundation for humus management until now. After years of work on its development, F. Sekera has managed to find a method for testing the soil's conditions for humus as well. The humus check is based on the assumption that:

1. the soil must be provided with sufficient quantities of organic material to serve as food for the small animals and microbes in the soil, and
2. the soil organisms must consume the food offered to them in a productive manner.

Accordingly, the humus test answers two questions:
1. How much food is available in the soil for the organisms?
2. To what extent is it consumed into humus?

To this end, a soil sample is taken from the field and the organic remnants of the decomposition are isolated and weighed. Any organic soil material that is already aggregated with the mineral components into a structural element of the soil is not captured in this process. The resulting organic material is then examined under a microscope. This allows you to study the juxtaposition of endogenous humus formation via the animals in the soil and exogenous humus formation via the local microflora.

The samples of organic matter first allow you to make qualitative comparisons between different soils based on the amount of material remaining. It's not until you see it through a microscope, however, that you can differentiate between the variety of types of humus and their different stages. You can see:
1. Undecomposed organic matter that is still fresh (root and stubble remains).
2. Extensively decomposed, mineralized organic matter without humus formation.
3. Organic matter that has been turned to humus to varying extents with a cell complex that is still visible (exogenous humus formation).
4. Fecal deposits from soil animals in the organic matter (endogenous humus formation).
5. Black humus aggregates without any recognizable tissue structure.
6. Small, cinder-like or carbonaceous parts.

Figure 36

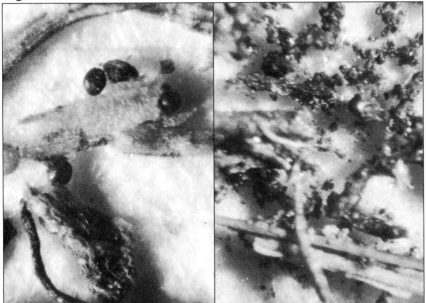

The formation of new humus in a field and its balk.
Left (field): the total amount of organic material accumulated is very low. It is fully decomposed and silicified, but no humus formation is evident. Scattered coprogenic humus is present. The result is residue from the harvest that breaks down quickly without building new humus.
Right (balk): the border is rich in organic material. Along with a large amount of coprogenic humus formation, everything from weak to strong exogenic humification is visible. The result is a sufficient supply of organic material and good formation of new humus.

In addition to the leftover organic material you will also find the dead bodies of individual small animals and numerous cysts. Keep in mind while assessing the samples that each one is just a snapshot of the decomposition processes taking place in the soil at a given time, meaning that you can't get much insight into what's happening in the soil until you've made multiple observations over time. However it can be helpful to take comparative samples from different locations, as long as they are taken at the same time. Investigating humus formation on one field and the balk surrounding it, for example, gave the image in Figure 36.

This method gives you an idea of organic sustenance in the soil. In practice, it allows you to assess any field where there are insufficient nutrients for the organisms in the soil or where new humus formation is too weak. You can then primarily focus your aid efforts there. As with checking the nutrients, the humus should be checked on a regular cycle, not just to verify whether measures taken during plowing were successful, but also because this kind of repeated observation is the best way to determine which strategies will result in the most productive field possible.

Chapter 8

CAUSES OF SOIL FATIGUE

Many expressions used in agriculture shift in meaning over time and only gradually acquire a consensus definition. Since time immemorial "soil fatigue" has referred to diminishing yields, even before anyone knew the details of the causes.

The phrase often referred to a depletion of the nutrients in the soil, sometimes to disruptions in the water supply, and other times to pest infestations that reduced yields. Today, if these issues are noticed they can be effectively fought against. But a plant's growth can also be stunted without any apparent reason, and you can't expect any lasting success in remedying the situation unless you can identify the causes. **You should look for the causes behind the disruption of the coexistence of the plant and organisms in the soil.**

It makes sense that the plant forms a sort of biotic community with the soil microorganisms since plants directly intrude upon the living space of soil microflora and microfauna. You have to think about it sociologically rather than considering each organism individually, as they are not capable of surviving separately but always depend on the society of other species. Today we know that plants form a biological organization (biocenosis) with the microflora and microfauna in the soil in which all members both contribute and receive something. We also know that this biological organization allows fertile soil to arise from dead rock and that even in hydroponics—which seems so contrary to nature—plants coexist with microorganisms of every type.

The coexistence of plants and soil organisms is most evident in the area immediately surrounding the roots: the rhizosphere. This area is densely settled by fungi and bacteria. The root is wrapped in a coat of organisms which makes the necessity of coexistence especially clear.

This coexistence of plants and soil organisms depends on reciprocity, or symbiosis. While roots continuously excrete organic material and shed cells, delivering food to the microorganisms, the organisms also help the plants by breaking down the nutrients into a usable form. It's very likely that plants and organisms also exchange materials whose particular functions we do not yet know.

This sort of symbiotic rhizosphere can be found in any unworked soil and in healthy fields (i.e., in soil that contains a diverse community of organisms in which bacteria, fungi, and animals produce the beneficial and detrimental materials characteristic of their respective species and keep each other in check through this elaborate network). It's impossible for any one member to proliferate on its own, a state of affairs accurately described as a biological equilibrium.

However in many fields you can find a completely different situation. The less diversified the field's crop rotation is, the less diverse the organisms living in the soil will be as well. You could think of

Figure 37

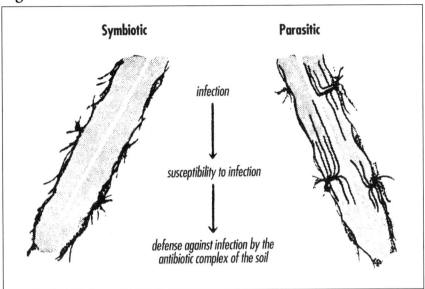

Schematic depiction of the symbiotic and parasitic rhizosphere.
Left: soil organisms settle on the root and live with it in a state of mutually productive symbiosis.
Right: soil organisms break into the root fabric and become parasites.

it as a collapse of the society in the soil that destroys the biological equilibrium. The severity of this collapse was already discussed in the chapter on humus, so we'll just provide a brief reminder here of how quickly the organisms in the soil decline when a field is plowed (Figure 31, page 73).

The progressive depletion of the community of organisms would be even more blatant if the small animals in the soil (worms, mites, springtails, etc.) also appeared in the experimental results.

How does this social collapse among the life in the soil affect the plants? To answer this question a schematic image of the rhizosphere is given in Figure 37.

The left side shows a state of symbiosis characterized by dense settlement of the root and its immediate surroundings by the soil organisms. In this state the root is able to provide the organisms with a steady supply of nutrients by continually shedding cells and excreting organic metabolic products. For their part the organisms aid in moving the nutrients around, which maintains the plant's nutrient supply. The image on the right, on the other hand, shows how things can degenerate into a state of parasitism if the organisms break into the root tissue and multiply there which disrupts the plant's water and air absorption.

This sort of fungal growth in the roots has been frequently noted before and was studied particularly extensively by A. Winter and his collaborators. Many authors, including A. Howard, believed that this phenomenon represented a beneficial symbiosis between plants and fungi, similar to the mycorrhizae that appear on trees in the forest. It is indeed thought that this is their initial function, but in many cases the situation in the rhizosphere causes the crop to degrade, at which point the relationship shifts from symbiosis to parasitism. Many examples in nature prove that there is no solid boundary between these two states, and many different types of transitions between the two can be observed.

Fungal growth in the roots is often merely temporary, especially during times of drought. Its appearance is accompanied by a general weakening of the root system which gives the impression that the plant's development is being stunted even though it shows no signs of sickness. This sort of fungal growth restricts the function of the plant's vascular system, meaning its absorption of water and nutrients. At any rate, the coexistence between the plant and the soil organisms is disrupted, and we have to conclude that we're dealing with a very widespread, previously unknown form of soil fatigue.

People first received information on the proliferation of this form of soil fatigue when F. Sekera succeeded in testing the soil with the dark germination experiment to determine whether the soil was forming symbiotic or parasitic rhizospheres. Starting from the fact that seeds do not germinate in the absence of light, you can observe different germination processes depending on the soil conditions in the germination medium. Either the dark germ cells develop normally and decay until the reserves within the seeds are used up after eight to ten days of autolysis or the seedlings succumb to a deadly infection and quickly break down during the period of longitudinal growth. Using a microscope you can perform more detailed analysis to determine whether an infection or autolysis is taking place. Numerous experiments with soil taken from grasslands, cultivated fields and balks, from both the topsoil and the subsoil, have proven that the cause of soil fatigue lies in a lack of variety in the crop rotation, as the most severe cases of parasitic rhizospheres occur in intensive operations that only grow beets, barley and wheat. It's true that limited nutrient uptake in the roots can be offset to a certain extent by adding an increased amount of fertilizer within certain limits, seemingly overcoming the soil fatigue, but nobody would argue that a healthy root couldn't make even better use of the fertilizer. Some crops react more sensitively than others to soil fatigue. Beets and barley are especially susceptible; rye and oats seem to be considerably more robust.

This leads to the question: **What prompts the soil organisms to become parasitic?** An infection is a natural consequence of dense organism settlement. Whether this infection penetrates the root tissue and spreads dangerously from there depends on how vulnerable to infection the plant is (i.e., whether the plant's living tissue is providing a strong supply of nutrients to the attacking germs). In a field amenable conditions occur repeatedly as every period of passive water balance and every surplus of soluble nitrogen (nitrogen surge) causes the root tissue to be enriched with amino acids and proteolysis products which serve as an excellent source of carbon and energy for the microorganisms. Whether an infection does make it through and spread in the tissue during a period of vulnerability depends in the end on the plant's resistance to infection. The plant is only capable of this resistance, however, if it is supplied with sufficient resistant materials (antibiotics) from the biocenosis in the soil. Healthy soil, with its diverse collection of life, is a natural "antibiotics factory," but this factory goes on strike if an undiversified crop cycle causes a decline in the population and diversity of soil inhabitants. In this case, the plant is no longer receiving the resistant materials from the soil and will fall easy victim to an infectious attack by the soil organisms.

Figure 38

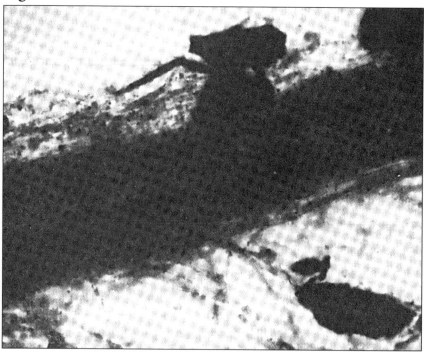

Rhizosphere of a seven-day-old cress root (100x magnification).

Figure 39

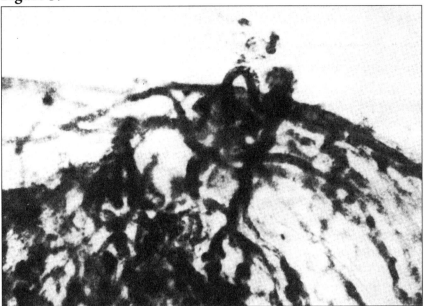

Parasitic rhizosphere of a three-day-old cress root. Fungal hyphae have broken into the root fabric (100x magnification).

How exactly a young seedling's rhizosphere expands, taking on either a symbiotic or a parasitic form according to the soil's resistance factor, is documented in the accompanying photomicrographs.

Figure 38 shows the symbiotic rhizosphere of a cress root. Germ growth in the soil from the balk around a field was observed for seven days without the root becoming infected. Then the seedling succumbed to autolysis.

In Figure 39 a cress root was observed in the soil of its field. The fungal hyphae penetrated the outer skin of the root after only three days; the soil did not contain enough resistant material to withstand the infection.

Figure 40 demonstrates the spread of the fungal hyphae in the sprout tissue (sprout base) of the same plant. You can clearly see how the vascular system is surrounded and penetrated.

Figure 40

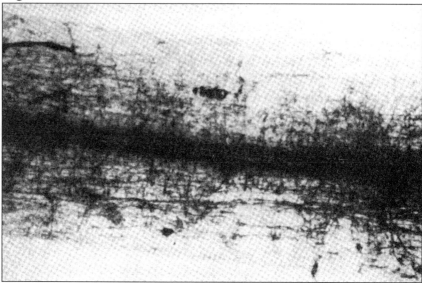

Parasitic rhizosphere of a garden cress after three days of sprouting with fungal hyphae grown through (100x magnification).

In the United States, people have supposedly successfully sprayed the missing resistant materials directly on the plant. But considering how closely field work already resembles a pharmacy, it's better to address these things directly at the root instead of hunting for fleeting success with partial solutions.

You have to get the "antibiotic factory" working again. You can do this even in spite of an undiversified, intensive crop cycle through the use of catch crops and cover crops during the fallow periods in the main crop cycle. This practice provides the soil with some welcome variety, especially if you make use of seed mixtures rather than just a single type. These measures at least give the soil a temporary biotic community. Mixtures of different plants have been used for a long time as catch crops, but a clover-grass mixture is a good start even as a cover crop. You can go a step further by combining living and dead plant cover in one of the many variants described earlier. The goal is to provide the soil with as much organic material in various stages of decomposition as possible in order to restore the organization of the life in the soil. This once again brings us back to the formation of communities of organisms (figure 32, page 75) and reemphasizes the fact that only a diverse biotic community can jump start a stagnant "antibiotic factory" and remedy soil fatigue. This is more than just a partial success as the same techniques can be used to attain excellent tilth and organized humus management.

We can't end a chapter on soil fatigue without a few words on the danger of nematodes, the cause of beet fatigue. Beet cultivation, which takes up a third of the field area in many intensive areas, sooner or later must lead to a dangerous infestation of sugarbeet nematodes (*Heterodera schachtii*) in the soil. This infestation can sometimes be so severe that beet production in an entire region will face an acute crisis. The problem has become more significant recently as intensive operations took an interest in enlivening their undiversified crop rotations and curbing decreasing friability in their fields with short-term rapeseed plots (Liho rapeseed and rapeseed cover crops). Producers were warned that sugar beet nematodes can live on any plant in the mustard family, and can even multiply in short-term rapeseed plots under certain conditions. In Austria, no cases of nematode infestations in rapeseed covers were detected in sixteen years of observation, but in 1954—a year characterized by abnormal weather—one was reported in northern Germany. This alarm prompted F. Sekera to examine the nematode problem from the perspective of soil biology.

The threatened farms are understandably seeking an effective nematacide, and the industry will presumably deliver one in the foreseeable future. A compound of that type would actually be of immeasurable value if it was able to radically "decontaminate" af-

Figure 41

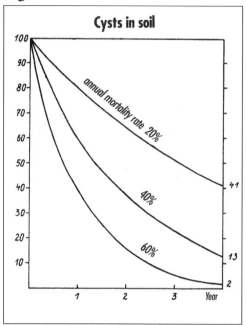

Cyst mortality

fected soil, and the farmer was determined afterward to relax his beet production with a reasonable crop rotation, preventing the nematode threat in the future. But if the future nematacide serves as a permanent crutch for beet farmers then it wouldn't just kill off the parasitic sugarbeet nematodes, but also the millions of free-living soil nematodes at the same time which would be an undesirable, destructive impediment to your humus management. Soil nematodes are among the valuable humus creators in the soil. Suppressing them means endangering the soil's fertility, and the consequences of taking a measure like that are totally unpredictable. In light of this concern it seems wise to approach the problem from another angle. Researchers have observed the life cycle of the pest and determined how many sugarbeet nematode cysts survive the transition from one host plant to the next (i.e., the mortality rate of nematodes in the cyst stage). Whereas useful soil nematodes breed through a constant turnover of generations, cysts of parasitic nematodes lying in the soil are deprived of their host plants after three to four years which also temporarily suspends their infectious attacks. The more active the soil, the more effectively the parasitic nematode population will be reduced. There is thus a causal relationship between the intensity of bacterial and fungal life in the soil on the one hand and the mortality rate on the other. This relationship makes a considerably simpler and more natural approach to fighting the pests possible. The significance of the cyst mortality rate can be gleaned from Figure 41 which calculates the drop in population of an initial group of a hundred cysts with various different mortality rates.

With an annual mortality rate of 20 percent, the population dropped to forty-one living cysts after four years. With a 40 percent mortality rate, however, it dropped to thirteen, and at 60 percent to two. These numbers show how inactive soil helps preserve the cysts while active soil can decontaminate itself, so to speak.

The fight against the nematodes thus far has indeed been moving in the same direction. Since time immemorial, adding legumes, clover or grass has been recommended as the most effective remedy to nematodes. This technique's success lies in the ability of these plants to increase soil activity which raises cyst mortality to a corresponding degree. This is a fundamentally biological approach to soil decontamination, even though some beet farmers might not see much appeal in such a tactic. Nowadays, the use of a cover crop can encourage activity in the soil effectively combating the nematode plague.

The following experimental results show that rapeseed cover crops are also capable of significantly increasing the cyst mortality rate in the soil. In 1954 there was a field of winter wheat moderately infested with nematodes that had had a rapeseed cover initiated in one section of the plot in fall 1953. This section was sown on August 14 and plowed up on November 8. The roots of the six-week-old rapeseed cover showed no signs of damage from nematodes. In early September 1954, a year after the rapeseed cover, the populations of living cysts per hundred grams of soil were tested.

Sample location	without rapeseed cover	with rapeseed cover
1	15.5	7.9
2	12.5	12.0
3	17.9	10.2
4	10.1	5.0
5	9.6	5.4
Average	13.1	8.2

These results imply that the cysts in the soil died off at a higher rate because the rapeseed cover improved the friability of the surface. Later results provided similar findings.

a. Determination of the structural conditions
without rapeseed cover
0–0.8 inches (0–2 centimeters) moderate surface compaction
0.8–5 inches (2–13 centimeters) thorough root coverage, friable
5–6.7 inches (13–17 centimeters) level 1 compaction
6.7–9.8 inches (17–25 centimeters) level 2 compaction
From 9.8 inches (25 centimeters) subsoil

with rapeseed cover
0–7.9 inches (0–20 centimeters) good root coverage, friable
7.9–9.8 inches (20–25 centimeters) level 1 compaction
From 9.8 inches (25 centimeters) subsoil

b. The amount of decomposed organic material (food for the organisms) per 3.5 ounces of soil at a depth of 0–7.9 inches (100 grams of soil at a depth of 0–20 centimeters) was:
- 0.23 grams with no rapeseed cover
- 0.31 grams with a rapeseed cover

The rapeseed cover significantly improved both the structural conditions of the soil for the subsequent wheat and the supply of food for the organisms living in it. Both of these benefits justify maintaining a healthy biotic community which has the secondary effect of attacking the nematode cysts more aggressively, causing them to die off at a faster rate.

In conclusion, we can say that the loss of the "old strength" of the soil is not only responsible for deficiencies in humus formation and friability but also is the likely culprit when the coexistence of plants with the organisms in the soil is disrupted. These problems do not appear in soil in nature because it is governed by a fine-tuned society of plants and microbes. Imitating these natural conditions in a field can only help to alleviate or avoid damage if the crop cycle is stimulated through the use of cleaning crops. A field that is always green is ideal from the perspective of soil fatigue as well.

Chapter 9

IS LIVESTOCK-FREE AGRICULTURE POSSIBLE WITHOUT ENDANGERING SOIL FERTILITY?

Animal husbandry and agriculture have grown together over the centuries to the point where they're now perceived as a single unit that arose organically, to the extent that farmers somewhat anxiously question whether one of the two components can ever be separated from the other without causing harm. And there are both operational and, more significantly, labor management and methodological reasons why this issue is becoming increasingly important in agricultural regions. The issue must be addressed first and foremost in the context of the needs of the soil because maintaining soil fertility is the most important goal of all. To prove the claim that it's entirely possible to manage a field without manure or fodder plants **if certain measures are taken**, we will have to find a practical solution that is both suitable from a biological point of view and operationally practical. Frequently, however, there's a lack of practical experience to draw from, especially regarding the latter point, so we will look for a solution that does not require revolutionary experiments and allows an organic development.

Our suggestion is to first separate operations into two zones. In the plots close to the farm carry out normal operations with livestock (Zone A) while making use of livestock-free agriculture in the outlying plots (Zone B). Once you've collected enough experience over the course of a few years you can shut down Zone A and use the crop rotation plan everywhere.

Zone A: The land area of this zone should be dictated by your feed requirements. The most important thing is to incorporate fodder plants into the crop cycle. The best option is alfalfa, which hasn't been in crop rotations in the past, used as a forage crop lasting for four years or more. This practice developed for operational reasons but now must be revised due to the desperate need for "good previous crops" in fairly monotonous crop rotations. As far as feed production goes, there's no dropoff in comparison with using it as a forage crop for four years, and yields in fact increase after the first year. But if you look at the alfalfa in terms of its value as a "previous crop," you'll discover that friability drops off noticeably by as soon as the third year and is completely unsatisfactory after four years. Alfalfa's ability to help increase friability is limited to the first and second years that it is used, a fact that should be exploited in the crop rotation. Alfalfa can thus only be used for two years at the most. In many cases, you'll need an alfalfa plot that lasts only one year due to the crop cycle not cycling around quickly enough.

We have observed significant positive results with this method for many years now, especially in dry areas. But you can go even further. If you categorize crops by their effect in increasing friability, you'd find the least friable soil under root and cereal crops and significantly more friable soil with legumes, other feed crops and grasses. The best possible performance in tilth and humus formation, however, comes from a clover-grass mixture. Because of this, a one-year alfalfa-grass crop (about 33 pounds or 15 kilograms of alfalfa and 18 pounds or 8 kilograms of orchardgrass per hectare) is preferable to using pure alfalfa and should be incorporated into the crop rotation. This would create a backbone for tilth and humus formation that the next crop could then draw on.

Depending on how much straw you need in the stable, the following crop rotations are options for Zone A:

a. Shallow straw (13 pounds or 6 kilograms of straw litter per livestock unit):
alfalfa-grass / root crop / spring cereal + rapeseed cover / winter cereal seeded with alfalfa-grass mixture.

b. Deep straw (26.5 pounds or 12 kilograms of straw litter per livestock unit):
alfalfa-grass / root crop / spring cereal + rapeseed cover,/ part winter cereal + rapeseed cover, / part root crop and feed corn / winter cereal seeded with alfalfa-grass mixture.

Manure is delivered along with the root crop, and the winter cereal is seeded with alfalfa-grass mixture.

If the alfalfa-grass mixture and the beet plot form the food source on their own, then you should plan on plot sizes of about 0.5 per livestock unit to be able to confidently cover the average yearly demand of 2.2 tons or 2 metric tons per livestock unit.

Zone B: Here you have to make do without manure or forage crops, which can lead to a number of issues.

The question is: **Can you replace the manure by plowing under straw, beet leaves, etc.?** We know that humus forms in the intestinal tracts of animals, but there's no law that says that it must be a cow's intestines specifically. Nowadays, with the secrets of the soil much more thoroughly researched, nobody would think to combine agriculture and animal husbandry simply to provide manure for the fields. The small animals living in the soil don't just create humus cheaper, they also do it better, as discussed in the previous chapter. You simply have to provide them with suitable living conditions. Straw doesn't need to be converted into humus on a manure pile if you can make this process possible in the soil itself. What matters is producing humus; how it's accomplished is of secondary importance.

The straw can only be converted into humus if the small animals living in the soil are made active, something best accomplished by planting a cover crop that can shade the soil and create the balanced soil climate that the animals need once the straw has accrued. We've already described the "clover-straw cover," but a rapeseed cover is also enough for cut-up straw to form an abundant amount of humus. With the use of samples of organic material from humus checks, researchers have shown repeatedly that straw fizzles out uselessly in the absence of a cover crop. With the protection of a cover crop, however, not only does humus form, but the straw is also broken down at a slow pace guaranteeing lasting soil activity. Now that that's been clarified we've created a foundation for livestock-free cultivation.

The cover crop isn't just needed to ensure humus formation—it also breaks up monotony in the crop cycle. Declining to grow feed means passing on the best possible preceding crops, so you have to find a substitute to keep soil fertility from deteriorating. This is true in general and particularly so for livestock-free agriculture, which tends strongly toward the undiversified cultivation of cereals. This sequence:

winter cereal + clover cover / spring cereal + rapeseed cover produces, from the perspective of the soil, a biologically flawless crop cycle of cereal/legume/cereal/mustard, providing tilth of a quality not attainable without a cover crop. If future developments make it possible to plant a rapeseed cover with a spring cereal by adding the seeds of a non-legume, then this will also do away with the period of intensive labor necessary to carry out stubble breakage immediately after the combine harvester.

One thing is certain: cover crops are very important alongside manure to eliminate the fallow periods that sap the soil's friability and humus supply, but they're absolutely crucial in livestock-free agriculture for the plowed-under straw to become humus and for the necessary variety in the crop cycle to be maintained.

Some details about how much organic matter is left behind by the individual stages of the crop cycle should provide insight into the soil's humus management. Although quantity isn't the only thing that matters, the importance of the nutritional value of the organisms' food means these figures are still very informative.

The current predominant crop cycle in non-livestock operations, beets/barley/wheat, delivers the following amounts of organic material, including the addition of 8 tons per acre (200 decitons per hectare) of manure.

Manure	1.5 tons per acre (34 decitons per hectare)
Beets	0.4 tons per acre (10 decitons per hectare)
Barley	0.8 tons per acre (17 decitons per hectare)
Wheat	1 tons per acre (23 decitons per hectare)
Total	3.7 tons per acre = 1.2 tons per acre per year (84 decitons per hectare = 28 decitons per hectare per year)

Based on past experience this is far too little to maintain the life in the soil. The crop cycle suggested for Zone A above, with 12 tons per acre (300 decitons per hectare) of manure added, delivers:

Manure	2.2 tons per acre (50 decitons per hectare)
Alfalfa-grass	1.8 tons per acre (40 decitons per hectare)
Beets	0.4 tons per acre (10 decitons per hectare)
Barley	0.8 tons per acre (17 decitons per hectare)
Rapeseed cover	0.9 tons per acre (20 decitons per hectare)
Wheat	1 tons per acre (23 decitons per hectare)
Total	7 tons per acre = 1.8 metric tons per acre per year (160 decitons per hectare = 40 decitons per hectare per year)

The improved crop cycle on the one hand and the concentration of the manure in a smaller area thus provides the field with 43 percent more organic matter!

In the livestock-free Zone B the constituents of the crop cycle deliver the following quantities of organic residue (including straw and leaves):

Beets	0.4 + 1.8 = 2.2 tons per acre (10 + 40 = 50 decitons per hectare)
Barley	0.8 + 1.1 = 1.9 tons per acre (17 + 25 = 42 decitons per hectare)
Rapeseed cover	0.9 tons per acre (20 decitons per hectare)
Wheat	1 + 1.8 = 2.8 tons per acre (23 + 40 = 63 decitons per hectare)
Clover cover	0.9 tons per acre (20 decitons per hectare)
Total	8.6 tons per acre = 28.7 tons per acre per year (195 decitons per hectare = 65 decitons per hectare per year)

Rapeseed cover	0.9 tons per acre (20 decitons per hectare)
Barley	0.8 + 1.1 = 1.9 tons per acre (17 + 25 = 42 decitons per hectare)
Rapeseed	0.4 + 1.5 = 1.9 tons per acre (10 + 35 = 45 decitons per hectare)
Rapeseed losses	0.4 tons per acre (10 decitons per hectare)
Wheat	1 + 1.8 = 2.8 tons per acre (23 + 40 = 63 decitons per hectare)
Clover cover	0.9 tons per acre (20 decitons per hectare)
Total	8.8 tons per acre = 3 tons per acre per year (200 decitons per hectare 67 decitons per hectare per year)

Wheat	1 + 1.8 = 2.8 tons per acre (23 + 40 = 63 decitons per hectare)
Clover cover	0.9 tons per acre (20 decitons per hectare
Barley	0.8 + 1.1 = 1.9 tons per acre (17 + 25 = 42 decitons per hectare)
Rapeseed cover	0.9 tons per acre (20 decitons per hectare)
Total	6.4 tons per acre = 3.2 tons per acre per year (145 decitons per hectare) = 72 decitons per hectare per year)

The values given here are very precisely calculated averages that include only the organic residue present at harvest time. They would be even higher if they included the root and plant matter shed by the plant and broken down during the growing season. In any case, the data makes it clear that livestock-free cultivation can provide the soil with an abundant supply of organic material and poses absolutely no threat to soil fertility. You just have to make certain that the organic material accruing in the field isn't squandered too quickly, but instead converted into humus with its long-lasting effects. You need cover crops to do that. Anyone who can master this technique will have success even without livestock. The decisive factor is having access to farm machinery that can accomplish all of these tasks without delays.

Chapter 10

FERTILIZATION: ONE FACTOR IN MAINTAINING SOIL FERTILITY

Fertilization and soil health always share a close relationship. One the one hand the soil's friability and humus status influence the fertilizer's effects, while the fertilizer, on the other hand, contributes to the maintenance of both statuses.

If you take a look at a cross-section of a non-friable field, you'll realize why the plants' nutrient supply can be a limiting factor. We think of the topsoil created by the plow as carrying out the storage and delivery of the nutrients, but this actually is only true of the portions of the upper topsoil that remain crumbly and high in root mass during the growing season. This active portion of the topsoil can sometimes become very shallow, and areas of topsoil compaction and the nutrient stores lying underneath then contribute almost nothing to the plants' sustenance because the roots cannot reach these areas. It can get to the point where plants are starving even in nutrient-rich soil because the topsoil is not being fully exploited as a store of nutrients. It's commonly believed that you can make up for this shortage and correct mistakes to a certain degree by adding extra fertilizer. But it's questionable whether this is a productive use of effort, as doing so can also lead to other dangers. The compaction cuts the plant off from the water reservoirs in the subsoil, which threatens its water supply. And without a supply of water nutrients cannot be exploited; the two go hand in hand. It is therefore not uncommon for fertilization to not only fail to increase yields, but to even do the opposite due to the additional nutrients which initially cause vibrant plant development and use up the limited water supply too quickly bringing things to the brink of disaster if it doesn't rain. It's thus important when giving advice about fertilizer to also provide instructions on how to make the fertilizer actually work in a field. It's easy to understand that the nutrition and water supply of plants are closely intertwined and that nutrients are therefore useless as long as the nutrient supply in the topsoil is separated from the water supply in the subsoil by a barrier layer. In other words, fertilizing is only worth the effort in healthy soil. Research institutes test thousands of soil samples for their nutrient content every year in order to give farmers basic guidelines for planning their fertilizer application. But these advisories depend on the assumption that the detected nutrients are actually available in the field. This is not the case in a non-friable field, which is something that should also be clarified.

This shouldn't be taken to mean that a farmer should reduce his fertilizer use if he's trying to make his field healthy. The foundation of friability is and remains manure. Using a shallow layer of properly decomposed manure in the course of crop rotation is an obvious step to any good farmer. Friability cannot replace fertilization; you have to give the soil back what you take from it at the harvest. There's a risk of losing too many nutrients, because while non-friable soil also uselessly immobilizes fertilizer nutrients to large degree, the vibrant activity of organisms in a friable field mobilizes them allowing plants to exploit them much more efficiently. Friable soil is fertilizer-active while non-friable soil is fertilizer-inactive. As with so many things in nature, the gears are connected here.

And now the second aspect of the fertilizer issue: **can mineral fertilizer help make soil more friable?** There have been repeated claims that artificial fertilizer is damaging and would kill the life in the soil and destroy its friability. Applying fertilizer inappropriately can certainly cause problems, but there are cases where even limited doses of (nitrogen-free) mineral fertilizer have had long-term beneficial effects. The warnings that it can destroy friability were indeed warranted back when people exclusively used sodium nitrate and sodium-containing potash salts. Since then we've been forced to recognize that mineral fertilizer in any form damages the life in the soil. For every nutrient that a plant needs there is some life form in the soil that will process it at will and under some conditions even transfer it to the plant roots. If easily soluble nutrients are added to the soil, these life forms will die and the soil system will be disturbed and become more vulnerable to harm. You will be able to detect these consequences the reduction of friable crumbs.

A healthy lime state in the soil is important. It's well known that most crops can't tolerate acidic soil. Some can manage a slightly acidic environment, but many require neutral to slightly basic conditions (pH 6.6–7).

Acidic soil poses three dangers to soil friability:
1. In acidic soil, the binding substances in the soil structure are not water resistant enough—the necessary aggregate formation does not take place, causing the soil to alternate between too silty and too compressed, depending on the whims of nature.

2. It's so hard for organisms to survive in acidic soil that they cannot accomplish sufficient biological tillage of the topsoil or humus formation. Acidity is always a sign of disruption to the life in the soil.
3. Plants cannot produce as much root material in acidic soil. This means that not enough "bacteria food" is ever produced, which also leads to deficient biological tillage.

As a rule, acidity indicates issues with the soil quality in the entire root area. The way to address it is by using large amounts of root mass and adding humus to stimulate a diverse range of life in the soil, which will then adjust the local pH to something close to neutral by itself. If it fails to do so, the cause may be a lack of calcium. Liming is a sensible option, but only in small quantities to avoid destroying the soil life.

Limestone powder, for example, is not easily soluble, unlike quicklime, so it needs carbon dioxide in order to effectively neutralize the topsoil. This means that it only works in active soil, since it is only in active soil that microbes produce carbon dioxide. Any fertilization should begin with checking calcium levels.

If there are still areas without phosphate or potash, they can be augmented in small amounts with fertilizer.

Potash only has a limited influence on the production of root matter. It takes an extreme level of potash deprivation to significantly limit root production in the soil, which leads to the food for the organisms accruing unusually slowly. In this kind of situation potash fertilization brings good results.

Phosphate, on the other hand, is extremely important to plants. Phosphate plays a special role by particularly helping to multiply the fine fibrils and root hairs that twist through the soil and supply food for the bacteria. Figure 42 shows rye roots in phosphate-poor and

Figure 42

Rye root (right) with phosphate fertilization and (left) without phosphate fertilization.

Figure 43

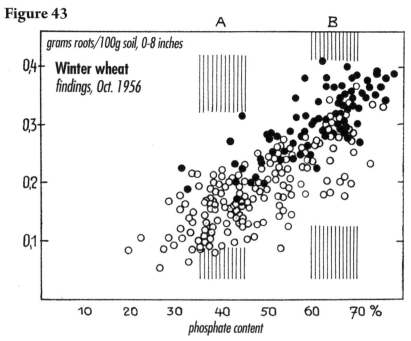

Root production in wheat relative to phosphate content and soil friability.

phosphate-rich soils, and you can clearly recognize the significantly increased numbers of the finest fibrils.

Figure 43 implies that wheat produces more root matter the better supplied the soil is with phosphate. The horizontal axis shows the soil's phosphate content, rising from left to right. The vertical axis shows the topsoil's organic matter content in grams per 100 grams of soil, rising from bottom to top.

In order to show the involvement of soil structure here, the cases where structural conditions were good are highlighted with a filled-in circle. A sufficient supply of organic matter for the organisms in the soil is clearly a prerequisite for supplying the plants with enough phosphate compounds and for producing friable soil.

Even the success of a cover crop is largely dependent on the addition of phosphates to the soil. The relationship between the proliferation of food for the organisms and the soil's phosphate content in a rapeseed cover crop was examined, and it's clear that higher phosphate levels and higher root production go hand in hand. Whereas the root mass remained under 0.05 grams per 100 grams of soil in extremely phosphorus-hungry soil, it increased to 0.08 grams per 100 grams of soil with moderate phosphate content and up to 0.12 grams per 100 grams of soil or more with a sufficient supply.

The individual aspects of soil life are intertwined like the gears of a clock here once again.

You can't measure the success of fertilization based solely on the increase in the next harvest's yields. This is another case where you must think in terms of the soil and come up with a long-term fertilization plan. Fertilizer's primary function is to encourage root growth. Its secondary function is to increase the plant mass on the surface which goes on to aid the life in the soil and humus production either directly as leftover material after the harvest or indirectly via livestock as manure.

This new perspective on the use of fertilizer corrects some of the errors of the past. It used to be said that fertilizer had a draining effect on humus; this was based on the fact that fertilized soil aspirates more intensively releasing more carbon dioxide. It's true that organic substances break down faster based on how nutrient rich the soil is, but you can't just consider one part of the phenomenon out of context. Stimulating the life in the soil increases root production (i.e., leads to better nutrition and increased multiplication for the soil organisms). Organic fertilization, which provides nutrients to the plant, also improves the living conditions of all the life in the soil thereby helping to maintain soil fertility.

An excess of nitrogen is dangerous not just to the plant, but also to the soil, because adding nitrogen on its own reduces the length of plant roots working against measures taken to increase friability. The simultaneous increase in the rate of root mass production usually balances out the drawbacks of the shorter roots.

Fertilization is an indispensable factor in maintaining soil fertility, but it can't work any miracles under poor conditions. A lack of friability in a field limits the success possible with fertilization. You can only achieve optimal results in a friable field. Fertilization is only a significant contributor to soil care if it helps to produce more food for the soil bacteria.

Chapter 11

WORKING THE SOIL IN A BIOLOGICALLY FRIENDLY MANNER

The most important device for working the soil is the plow, so it's no surprise that there is constant debate over what type of plow is the best. There's no dogmatic answer to this question because the benefits of one plow over another can only be decided on a case-by-case basis. What's isn't disputed is that the best plowing is plowing that gives the plant roots the best chance to develop, that creates friable topsoil with access to the subsoil so that the roots can grow deep and have access to the water reserves in dry periods.

But the plow is not just a tool for loosening soil. It also deposits the material left over from the previous harvest to where it can be used as food for the organisms the next year.

The question always comes up: **How deep should the topsoil be?** There's no single figure either in general or for a given specific field. It depends on how deep the plants can make the soil friable (i.e., how deep the topsoil's biological tillage reaches). J. Görbing believed that there are natural limits to where the soil can be made friable, a maximum of 8.6 inches (22 centimeters) in moderate climates and just 6.2 inches (16 centimeters) in hot areas. Later research by F. Sekera showed that these depths were underestimated. The depths also change with the seasons. Up to 11.8 inches (30 centimeters) or more can be made friable in healthy soil in the spring and fall with a shallower area in the hot season. This is because the more intensively the soil organisms aspirate, the more oxygen they need, which limits the range of soil layers that can provide a sufficient supply. You can thus never specify an exact point where the soil can no longer be made friable, just a vague range with fluctuations from season to season.

There are three big plowing mistakes that can ruin even the best soil:
1. The plow furrow is too shallow and always the same depth every year.
2. The soil is turned too deeply.
3. The plow furrow is too wide.

The first mistake is occurring less and less frequently nowadays. It used to be common on small farms. If you only turn the soil shallowly each year and always drag the plowshare (often even a blunt one) along the soil at the same depth, then you'll gradually create a hard plow pan underneath the shallow layer of crumbly topsoil that roots cannot break through. Water also collects on the plow pan, and plants will alternate between being too wet and too dry if the connection to the subsoil is broken. The space available to the roots is too limited for you to be able to expect a satisfactory harvest. Furthermore, plow pans make the perfect source of nutrients for couch grass. Couch grass's rhizomes form especially well in moist layers. They can't penetrate through the compacted plow pan so they form a dense tangle on the lower border of the topsoil and survive at the expense of the crop in the already limited root area. Figure 44 shows the penetration and spread of the couch grass roots at this border.

People learned from these mistakes and started to use deeper topsoil layers. That always causes an initial increase in yields. Available land is determined not by surface area but by surface area times the depth of the usable soil, and the plant has large reservoirs of water and nutrients now available. Nevertheless, you have to be careful not to make any mistakes when deepening the topsoil or your success could turn into a failure leaving you with greater difficulties than you had before.

Figure 44

Wildrye root spreading through a water buildup caused by soil compaction.

Turning too deeply is one such mistake. Doing so dilutes the densely rooted part of the topsoil with the sparsely rooted lower topsoil. As has already been addressed repeatedly, the soil requires an abundant supply of residue from organisms for biological tillage to create a crumb structure. You can tell from a topsoil's cross-section how the previous crop's roots spread through the soil. The upper portion is densely rooted, whereas the lower part of the topsoil has only sparse roots. This is significant because it tells you where the nutrients for the next year's organisms are located. Only the densely rooted portion of the soil can become friable because it's the only area that provides the necessary conditions for thorough biological tillage to take place. Until recently the rule of thumb was to turn cereal crops more shallowly and root crops more deeply. This would generally also encompass the sparsely rooted layer, meaning that people were burying the densely rooted soil that could become friable underneath the sparsely rooted soil that could not. This is the wrong approach because the stable crumbs need to be on the surface where they can be hit by rain. If they're not, not enough biological tillage will take place in the crumbs created by the plow on the surface, and they'll break up and turn to silt after the first rainfall. A crust will form preventing air from entering the soil. The lower portion of the topsoil would indeed provide excellent conditions for sufficient biological tillage in the crumbs since it would contain large amounts of root remains to feed soil organisms, but the crust would prevent the soil from being sufficiently aerated for this to take place. Biological tillage would therefore not even happen in the interior of the topsoil, leading to the topsoil gradually crumpling and suffering progressive disturbances to its cross-section, as already explained on page 32.

But even worse: the damage continues and has an even greater effect the next year. If you think about how the plow brings the compacted areas upward and brings clods to the surface, then it's easy to image what the consequences would be. It is possible, with great effort, to break up these clods, and the frost can loosen them, but you're missing the key factor: the crumbs are not stable! A one-time issue with soil friability can be accelerated and worsened each year with improper plow work.

What is the right way to turn the soil? Once you've used the spade test to figure out how deep the densely rooted layer is you can adjust the plow based on your findings. This layer is enough, even if it's shallow, as long as there are no compacted areas underneath it and the roots can grow downward unhindered. If there are compacted areas, they mustn't be brought toward the surface, but first loosened to the point that roots can penetrate them and bring new life to the compacted soil. How deeply you have to loosen the soil depends on the depth of the compacted area.

The right way to approach depth in tillage is therefore to turn the shallow soil and loosen the deep soil. This is best accomplished with a two-layer plow. A number of plow manufacturers offer two-layer plows of various designs from livestock-drawn plows with a single blade to heavy tractor plows. The principle lies in having a digging body that can be adjusted to different depths behind the normal plow body that turns the upper topsoil but loosens the lower topsoil without mixing the two layers and diluting the bacteria food as happens with deep turning. These digging bodies are usually falsely called "subsoil looseners." Loosening the subsoil is only necessary in certain special cases. The job of the digging body is generally limited to breaking through the plow pan and breaking up the compacted lower topsoil.

Setting up a two-layer plow requires a certain degree of caution. Most importantly, you must adjust the turning blade so that it reaches as deep as the dense roots of the previous crop in your field; the digging blade should be adjusted so that it reaches a little deeper than the lower topsoil which it should lift up and redeposit in a looser form. So there's no rigid formula for figuring out the depth; you have to base your decision on the conditions of a given field. This is the only way you can expect success and a hitch-free process.

There are essentially two different types of two-layer plows, those with digging bodies that cut and those with digging bodies that break things up. The latter variety has the advantage of not creating a sharp divide between the turned layer and the dug layer which gets you closer to the goal of unimpeded access between the topsoil and the subsoil.

However there's no point in using a two-layer plow unless it's immediately followed by a catch crop or a cover crop that increases friability (i.e., unless nutrients for the soil organisms are added into the mechanically broken-up topsoil so that biological tillage takes place). Even if the two-layer plow is the best and surest type for producing good tilth, it's still not able to induce biological tillage. **The plow and the plants must work together.**

The third mistake that you can make in plowing is excessively wide furrows. It seems an obvious move to increase the width of the furrows as well as the depth so that the plow could cover more area. But furrows that are too wide pose a serious danger of their own to soil quality. If you don't have time to wait for ideal moisture conditions in the soil, you have to use smaller furrows making sure that the plow is crumbling the soil and not breaking off one big chunk after another. Unfortunately, however, people are very quick to ignore this fact and just hope for frost to break up the chunks and clods.

Of course it's possible to rely on laborious post-treatment to change the surface of even a clod-filled field and obtain the fine earth necessary for cultivation. It's also true that the frost breaks the chunks open, but these effects are limited to the upper layer of the

Figure 45

Row of beets taken out of the ground showing the effects of clods from the plow.

topsoil. The interior of a field that has been plowed with furrows that are too wide remains as raw and clod-heavy as it was right after plowing.

There have been extensive investigations into how long these clods stay intact. They survive the entire growing season and up through the next time the soil is plowed! Figure 45 shows the results of experiments of this type. When the soil was dug up immediately prior to the beet harvest you could still see the intact clods. You could even determine exactly how deep and wide the plow furrows had been from their dimensions. The clods have grown together with the subsoil to a degree, or they've been broken up some by the blade and then lie embedded in the topsoil. In any case they represent a significant disturbance to the topsoil's cross-section and thus are also a danger to the harvest. You can clearly see how the beet roots avoided the clods. This wasn't simply because they were too hard, but also because the dense soil is not sufficiently aerated.

Working the Soil in a Biologically Friendly Manner

Figure 46

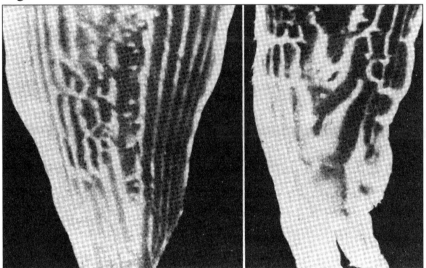

A rooted beet has its vascular system on the inside.
Left: healthy beet, right: rooted beet.

If a beet happens to be located between two clods, it will do reasonably well. But if a beet sits on top of a clod, it will remain weak and rooted. No beet resembles any other, which can be seen in the development of their leaves alone. This dissimilarity is an unmistakable sign that a field was plowed with furrows that were too wide and that there are still clods in the soil. It's no surprise that the rootedness of the beet also affects the interior of the beet body resulting in a degeneration of its structure. Figure 46 shows thinly cut beet slices.

In the healthy beet (left), you can make out a centralized vascular bundle, whereas in the rooted beet (right), there are connections between the bundles. This is how the beet—being disrupted by the clod—is trying to maintain its fluid circulation, but it requires an expenditure of energy that is then lost from the harvest. These sorts of vascular bundles also cause significant difficulties in processing the beets.

As mentioned before, there's no rigid formula for how to set up a plow. Different soils can require different things. This also makes it impossible to manage with just one plow body. You definitely need at least two types: one that turns fully and one that just shifts the clods from the plow around without substantially mixing them. As long as the field is not in good tilth (i.e., topsoil compaction forms every year) you will certainly need a sheer body so that you don't bury the friable upper topsoil. It's certainly no coincidence that the animal-drawn plows that you can still find in barnyard scrap heaps generally did a better job of this than modern tractor plows. It may be because farmers used to walk in front of the plow but now sit on the tractor and don't pay as much attention to what the plow is doing. Take care to use a sharp blade as they can cut up the bottoms of the furrows better and help prevent a cohesive plow pan from forming. Even better would be blades that don't cut in a straight line but instead have harrow-like tines that could roughen the soil the way that harrows do.

The decision of when to do your plowing is also important. In general the goal is an orderly fall plowing. But there will always be some years when you don't manage it which brings up the question of whether to plow the field wet or to wait for spring. People who advocate absolutely always plowing in the fall, even if the furrows have to be "greased," always bring up the argument that frost crumbles the furrows. But that's only true on the surface; in the interior the clods from the plow remain in the same state as when they were first put there. Mistakes like this can often ruin a field for years on end, which isn't exactly a beneficial outcome.

Other farmers deftly make use of the pause in the winter to achieve a successful plowing without greasing the soil. These cases of "cold plowing" were carefully examined, and it was observed that winter furrows of this sort are especially vulnerable to silting and compaction, and often even collapse and form a crust at the first thaw. Any physical working represents a wound to the soil's structure and breaks up the organic connective material built by the organisms in the soil to solidify the crumbs created by the plow. These wounds have to heal every time that the field is worked; soil organisms have to build new connections, but can't if it is too cold and the field is lying torpid. The wounds thus remain open, and the next rainfall and the thaw are certain to cause the field to break down. This is also the reason you shouldn't work with a series of different pieces of equipment driving over the field with a new machine every few days but should instead use all necessary equipment in one go. A tractor makes this possible with its energy reserves.

This "cold plowing" is poison to the soil, so the remaining option—assuming you didn't manage to finish fall plowing on time—is the much maligned spring plowing. The criticism is at least justified when it comes to dry areas since turning the soil always leads to water losses. But the idea of a spring plowing is less repellant than a wet fall plowing or a cold winter plowing. There's one condition, however: the spring furrows must be made seed-ready in just one day. If the field sits in a raw state for even just a few hours after plowing the water losses can reach quite noticeable levels which makes it harder for the seeds to sprout. You will certainly want to use a roller harrow or star harrow together with the plow to immediately break any chunks into fine pieces, and this work needs to be carried out immediately after the plowing to make the field seed-ready. This isn't an attempt to make a case for spring plowing—it has always been and remains an emergency measure—but it's definitely preferable to an abortive attempt in the fall or winter.

Even the most modern plow cannot create ideal seed furrows unless they were prepared first with proper skim furrows. This fact is unfortunately still not acknowledged enough because the "stubble breakage" often fails to meet the demand set on it. The need to rush has often led to a lack of caution. If the stubble plow moves clods on top of each other and they then dry out, it's likely that the entire topsoil will dry out and harden such that the plow cannot dig in the seeds beds. The situation is not improved by just using deeper skim furrows to create a better track for the plow. The goal of stubble breakage is to create the ideal seed bed for the weeds in the field. The more weed seeds you can induce to sprout using the skim furrows, the less you'll have to destroy later. The more root matter you introduce into the soil, the more the topsoil is loosened and prepared for the seed furrows. You can only reach this goal, however, if a crumb cover forms during the stubble breakage with the stubble remains worked in very shallowly. This is another situation where you have to make use of the crumb cover principle! Then you must decide what equipment to use. A stubble plow usually only works in soft soils that pour well and don't form clods. A disk harrow seems better suited to hard soils. It not only allows you to cover more area but also does better work, especially if it's driven crosswise over the field. In any case you must be sure not to lose any time, because if the broken stubble is allowed to dry out once and lie there in a hardened state, the disk harrow won't be able to penetrate it any more. To press on the initially still very loose crumb cover and cause the weed seeds to germinate faster, a rotary harrow or roller harrow should also be used the second time you drive across. This post-treatment is absolutely necessary after using a skim plow. The field must not sit in its raw state; it needs a crumb cover so that the weeds sprout and the field can begin to be made friable. A disk harrow is not the ideal machine for stubble breaking, however, because it doesn't do a good enough job of clinging to the uneven parts of the field and thus doesn't do its work in some places.

Everything said about stubble breaking here also goes for the use of a rapeseed cover. The disk harrow is better suited than the skim plow to this task as well. If you break up a rapeseed cover the first time you drive over it with a disk harrow, then wait a few days until the leaves have wilted before driving over it again crosswise, you'll come out of it with a serviceable organic crumb cover. The faster you drive, the better it works. If you use 4-inch (10-centimeter disks), most of the organic matter will be located in the 2.4-inch-thick (6-centimeter-thick) crumb layer at the top, as is desired.

The cultivator is a practically indispensible piece of equipment for soil care because it can loosen the topsoil multiple times without mixing it together and causing it to lose its moisture. If you grow a cereal crop after potatoes, for example, you can usually dispense with plow furrows and reach your goal more quickly with a cultivator with an attached bar float. A cultivator is also necessary in spring if the seed furrows created in the fall have been too greatly diminished under the pressure of snow and you want to loosen things up again.

In terms of lowering the amount of effort you need to put in to work a field, a friable field is always much better prepared for any equipment. There's no need to fight against the clods because friable soil crumbles more easily, and in many cases you can make a field seed-ready in one pass. This is also a considerable timesaver because it avoids the need for laborious post-treatment. Some comparative figures on energy expenditure may be of interest. They show the differences in friable and non-friable soil. In the following table, energy expenditure is given in pounds per square inch (psi, or kilograms per square centimeter) necessary to push a steel probe 8 inches (20 centimeters) into the field.

Energy Needed in psi (kg/cm²)

Field Being Tested	Non-Friable Topsoil	Friable Topsoil	% Saved
Reichersberg	234.5 (16.5)	113.7 (8.0)	50.4
Lambach	217.7 (15.0)	127.9 (9.0)	40.0
Kröllendorf B.	199.0 (14.0)	106.6 (7.5)	46.5
Kröllendorf W.	270.0 (19.0)	127.9 (9.0)	52.7
Kröllendorf I.	312.6 (22.0)	170.5 (12.0)	45.4
Spillern 3	255.8 (18.0)	163.4 (11.5)	36.2
		Average	44.8

An issue that's stimulated a lot of discussion recently and causes much uncertainty in practice is the pressure field machinery exerts on the soil, especially with tractors. Sounding alarms over this is not very helpful. Dealing with this "tractor pressure" is not solely the job of the designer. The manufacturer's job is to make the contact pressure of the wheels as low as possible. The farmer can do a lot toward remedying this problem by increasing the field's load-bearing capacity and making it less sensitive to the pressure from the wheels.

What makes the tracks of the tractor wheels any different from the pressure caused by other farm equipment and treads? If you measure the waterlogging in the tracks caused by the pressure on the soil, you get the results given in Figure 47.

Figure 47 compares the tracks of a tractor to those of a trailer loaded to provide an equal amount of pressure. The motor-driven wheels of the tractor caused a serious deformation to appear in the upper area of the tracks (the "slippage area"), while the actual effects of the pressure reached deeper but were significantly milder and subsided as you went deeper. The wheel tracks of the trailer loaded to provide the same pressure on the soil were characterized solely by the pressure effects; there were no slippage areas with heavy deformation.

Figure 47

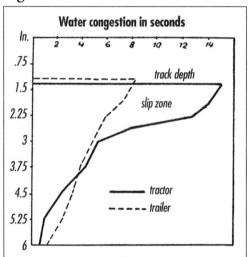

Water congestion in the tracks of a tractor and those of an equally heavy trailer.

We can conclude that there's a fundamental difference in how motor-driven wheels and wheels that are just rolling affect the soil. The damaging effect of the tractor wheels clearly comes in the badly deformed slippage area of the tracks. It should be added that this slippage area will form even if the wheels aren't visibly slipping themselves. The effect is greater the wetter the soil and the more powerful the tractor.

Based on the experience that friable soil does a better job of cushioning treads than non-friable soil, it was to be expected that the former would also be less sensitive to the pressure created by tractor wheels. This is true if you're only considering the effects of the pure downward wheel pressure. In the face of the grinding effect of the tractor on the upper slippage area, however, the biological tillage in the topsoil is insufficient. If, however, a friable field is able to withstand significantly more pressure and take less damage from the tractor than non-friable soil, this important distinction lies simply in the fact that a friable field can dry out more quickly after a rainfall and thus be driven on sooner. Figure 48 shows the results of an experiment that recorded the depth and thickness of the slippage area created by the cultivator in the tracks.

While three days of rainfall severely disrupted the non-friable portion of the field in question, and the soil only haltingly dried out afterward so that tracks were still distinctly visible after six days, the friable part of the field could be driven on again just two days later without incurring any damage. This distinction reveals the core of the whole issue. Friable topsoil always creates open access to the subsoil allowing the water gathered in the topsoil to be quickly distributed and the field to dry out again soon afterward. In a non-friable field, on the other hand, water gathers above the compacted areas of the topsoil and the soil only slowly dries. As water is uselessly lost via evaporation the tractor must sit inactive before returning to the field. The "tractor pressure" is largely dependent on the condition of the field. Maintaining a connection between the topsoil and the subsoil is critical as increasing the strength of this connection means a field that can withstand more pressure and less damage from the tractor's wheel pressure. Just as rain can become the enemy of a non-friable field and make it silty instead of reinvigorating it, a less friable field is more sensitive to the effects of tractor tire pressure.

Freshly loosened soil is particularly sensitive. As farmers moved more and more toward mechanizing not just the hard field labor but

Figure 48

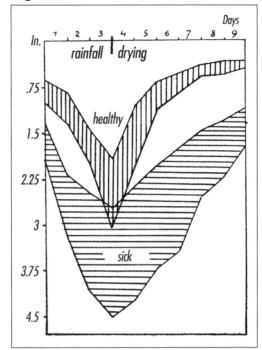

Slippage areas caused by the tracks of a cultivator in healthy and sick soil.

also the subsequent measures taken to care for the soil a problem arose. Of the different track looseners developed to loosen the track areas back up, none met with more than partial success. It wasn't until the appearance of lattice wheels, which distribute the pressure over a significantly greater area and leave crumbly soil between the areas that contact the surface, that a real tool against this kind of damage was available. Pressure tracks mellow much faster if they are covered with crumbly soil.

The preceding section was only able to address the most serious issues connected with working the soil. The key point to get across is that technical field work can be carried out more easily and with more certainty if you recognize the biological work that is happening. Again and again we return to the idea that plants are the foremost creators of good tilth, and that the best that can be expected from a mechanical method is a mechanical preparation for the ensuing biological processes. The more care you take to maintain your soil's friability, the easier and cheaper mechanical field work will become.

Closing Remarks

This little book provides a general plan for the soil and shows where disruptions to this organization stem from and how they can be triggered. We've also discussed how to combat this damage which can be categorized into:
1. Disruptions caused by nature (e.g., deliming and loamification) that age and lower the value of the soil.
2. Disruptions caused by cultivation (e.g., loss of friability or humus) that move the soil away from a state of nature.

We focused primarily on the latter category including consideration of the technical, operational and economic factors involved.

Soil fertility cannot be maintained by expending more effort, but only by meeting the natural demands of the soil.

The soil is neither a lifeless, weathered crust of stone nor a single organism, but rather a biotic community of plants, microorganisms and lower animals. As with any organism their individual biological, chemical and physical processes are inseparably intertwined. If one piece of this chain fails it demands an adjustment by the entire community just as how an animal or plant would be disrupted by one of its organs not functioning properly. There is no single organism that can support itself completely on its own. You need to think of an intestinal tract with its complex bacterial flora or of the inseparable coexistence between plant roots and the microbes in their rhizospheres. There is no known type of organism in nature that exists in isolation, organized communities. And any disruption to this organization must inevitably reduce its efficiency.

This is the reason for using the terms "healthy" and "sick" to refer to soil. Not in the figurative sense, like how you might talk about a healthy or sick economy, but in the purest sense of the words as they're used with people, animals and plants.

The methods of organized soil management also stem from this mindset. Soil biology involves researching the biological relationships among life in the soil; soil hygiene involves developing measures to use to maintain soil health.

Afterword

This book has its origins in the work of the same title by the author's husband, Professor Franz Sekera, an Austrian soil scientist who died in 1955. It was first published in 1943 by Reichsnährstands-Verlag, Vienna.

Index

Acidic soil, 114–115
Actinomycetes, 9, 86–87
Aeration, of soil, 2
Aggregate structure, of topsoil
    in acidic soil, 114
    consistency of, 9–10
    dissolution of, 14–15
    effect of water on, 9
    formation of, 3–4
    water resistance of, 14
    wind-blown, 26
Alfalfa, 105
Alfalfa-grass mixture, 105–106
Amino acids, 95–96
Ammonium nitrogen, 53
Animals, in soil
    earthworms, 76, 78, 80, 81
    humus production by, 76, 106–107
    insect larvae, 76, 78–79
    meal worms, 78–79
    nematodes, 99–103
    in plowed fields, 93
Autolysis, 95, 97
Azotobacters, 11, 12

Bacteria
    azotobacters, 11, 12
    weight per acre, 9–10
Barley
    in livestock-free agriculture, 109–110
    soil fatigue susceptibility of, 95
    soil structural deterioration and, 33–35
Beet fatigue, 99
Beets
    in clod-filled soil, 125–126
    as livestock feed, 106

in livestock-free agriculture, 108, 109
soil fatigue susceptibility of, 95
spade test for, 69–70
vascular system of, 126

Biocenosis
definition of, 16, 92
as resistance factor source, 96

Biological equilibrium, 92

Biological tillage, 3–4
in acidic soil, 115
conditions required for, 14–15
in crumb layer of soil, 10–14, 122
depth of, 119–120
as microerosion deterrent, 23
microorganisms' role in, 9–19

Biotic communities
buildup of, 74, 75
in cultivated *versus* uncultivated fields, 73–74
humus formation by, 76–78
interrelationships within, 135–136
plants as components of, 91–92
in soil crumb layer, 82–83

Black medick, 51, 78–79

Calcium, deficiency of, 115
Canola, as cover crop, 46, 50
Carbon, 95–96
Carbon dioxide, 16, 52, 78, 115, 118
Carbon turnover, 72, 79–80
Catch crops, 50
for soil fatigue control, 98
two-layer plowing and, 124

Cereal crops
cover crop for, 50–51
effect on soil friability, 105
fusariosis in, 64
interplanted with clover seeds, 33–35
optimal root depth of, 58–59
root networks of, 37
soil structure breakdown in, 30–35
turning of, 122

Chaff dung, 84–85
Chalk fertilization, 14
Clay-humus complexes, 74, 77, 86
Clods, in soil, 124–126, 127

Clover
as cover crop, 50–52
in livestock-free agriculture, 109–110
effect on soil structural breakdown, 33–35
for nematode control, 101

Clover fatigue, 51
Clover-grass mixtures, 105–106
Clover-straw cover, 52, 107
Colloidal chemistry process, in soil, 3–4, 14–15
Compacted soil, 16–17
    depth of, 123
    nutrient content of, 112–113
    plowing of, 123
    water balance/reservoirs in, 60–62, 64
    water flow rate in, 23
Compaction, of soil, 22
    adverse effects on plants, 36–37
    effect on roots, 34–35
    effect on taproot plants, 37–39
    level 1, 68, 69, 70
    level 2, 68, 70
    level 3, 68, 70
    process of, 21, 27–29
    spade test of. *See* Spade test
    timing of onset of, 30–35
Composting, 86–87
Couch grass, in compacted soil, 120–121
Cover crops
    canola, 46, 50
    for cereal crops, 50–51
    clover, 50–52, 109–110
    combined with manure, 52–53
    dew condensation in, 65
    disk harrowing of, 130
    effect on soil microorganisms, 74, 75
    in livestock-free agriculture, 107–108, 109–110
    mustard plants, 50
    for nematode control, 101–103
    as organic soil crumb layer, 83
    phosphate fertilization of, 117
    rapeseed, 46–50, 65, 74, 75, 101–103, 107–108, 109–110, 130
    for soil fatigue control, 98
    two-layer plowing and, 124
Cress roots, rhizospheres of, 96, 97–98
Critical water content, 63–64
Crop rotation
    lack of variety in, 92, 95
    in livestock-free agriculture, 105
    manure use with, 113
    root matter production in, 44–46
Crumb structure, of soil. *See also* Aggregate structure, of soil
    breakdown of, 5–7
    compaction of, 27–28

effect of biological tillage on, 10–14
effect of plowing on, 122–123
effect of rain on, 13–14
effect of stubble breaking on, 129
effect on plants' growing conditions, 19
optimal, 2–3
organic, 81, 82–83, 85, 86
   formation of, 81, 82
   in humus management, 82–83
   manure in, 85
   rapeseed cover crop and, 130
relationship to soil quality, 1–2
Crust, on soil, 27, 28–29, 122
   in winter-plowed furrows, 127–128
Cultivators, 130, 132, 133
Cysts, in soil, 89
   of nematodes, 100–103

Dark germination experiment, 94–95
Decomposition products
   in humus test samples, 89
   of plants, 76, 77–78, 81, 98–99
   for soil fatigue control, 98–99
Dew, 65
Disease and pest resistance, effect of water balance on, 64–65
Disk harrows, 129, 130
Drifting, of soil, 26
Drought
   compacted soil's susceptibility to, 29
   fungal growth during, 94
   irrigation systems and, 62–63
   plants' susceptibility to, 36
Drought-resistance, of plants, 63–64
Dry rot, 64

Earthworms
   decomposition of, 78
   effect on soil structure, 81
   feeding behavior of, 80
   role in decomposition process, 76
Energy expenditure, 130, 131
Erosion
   control of, 9, 10–11, 25–26
   furrow-type, 24
   macroerosion, 23–24, 25
   microerosion, 20–23, 24–25
   primary cause of, 24–25
   sheet-type, 24, 25

Europe, erosion problem in, 23–24
Evaporation, 29, 61, 62, 133
    of irrigation water, 63

Faulkner, H., 86
Fertilization/fertilizers, 112-118. *See also* Manure
    artificial, 114
    in compacted topsoil, 36, 37
    functions of, 117
    gypsum, 4, 14
    lime, 4, 14, 54, 114, 115
    mineral, 114–118
    in non-friable soil, 112–114
    "nutrient piles" in, 87
    organic, 118
    phosphate, 115–117
    phosphorus, 46–47, 115–117
    potash, 46–47, 114, 115
    relationship to friability, 113–114, 118
    sodium nitrate, 114
    water reservoirs and, 113
Flaig, W., 76
Fodder plants, 105
Franz, H., 76, 80–81
Friability
    depth of, 119–120
    fundamental conditions for, 14–15
    goal of, 57
    improvement of
        based on physical *versus* biological methods, 49–50, 51
        with cover crops, 46–53, 55
        plan for, 54–55
        with plowing, 53–54
        time requirements for, 54
    lack of, as soil erosion cause, 25
    loss of
        definition of, 4
        plowing-related, 56
    relationship to crop yield, 41–42
    relationship to fertilization, 13–114, 118
    spade test in, 54
Friable soil, 1
    comparison with compacted, non-friable soil, 4–5
    rain absorption rate in, 24
    root networks in, 37
Frost, 15, 65, 124, 127
Fungi
    actinomycetes, 9, 86–87
    growth on rhizosphere, 92, 94–95, 97–98
Furrows
    compaction in, 28, 29
    erosion in, 24

skim, 128–129, 130
stubble, 83
too deep, 120, 121–122
too wide, 69–70, 120, 124–126
Fusariosis, 64

Görbing, J., 27, 66, 67, 119–120
Grasses
  effect on soil friability, 105
  for nematode control, 101
Groundwater soils, 57–58
Gypsum fertilization, 4, 14

Hardpan, 56
Heart rot, 64
Howard, A., 94
Humus
  coprogenic, 89
  effect of fertilization on, 118
  effect on soil compaction, 17
  formation of, 16. *See also* Humus management
  endogenous, 77–78, 88
  exogenous, 77–78, 88
  friable, 72–74, 76
  microscopic examination of, 88–90
  nutrient content of, 87
  as plants' nutrient source, 87
  as root channel coating, 44
  soil-ready, 86–87
  stable, 72, 74–76, 86–87
Humus management, 72–90
  in acidic soil, 115
  carbon turnover in, 79–80
  components of, 72
  definition of, 72
  humus formation process in, 78–80
  humus testing in, 87–90
  mixed nutrients in, 78–80
  organic soil crumb layer in, 81, 82–83, 85, 86
  soil fatigue control methods in, 99
Hydroponics, 92

Infection, plants' susceptibility to, 95–96
"In good tilth," 1, 14
Insect larvae, 76, 78–79
Intercropping, 46–56
Irrigation systems, 62–63

Kidney vetch, as cover crop, 51
Kubiena, W., 76

Laatsch, W., 76, 86, 87
Lang, R., 10
Legumes
  effect on soil friability, 105
  for nematode control, 101

Lime fertilization, 4, 14, 114, 115
    effect on soil structure, 54
Limestone powder, 115
Livestock-free agriculture, 104–111
    cover crops in, 106, 107–111
    crop cycles in, 107–111
"Loam crests," 24

Macroerosion
    definition of, 25
    prevalence of, 23–24
Maiwald, K., 81
Manninger, A., 86
Manure
    application and placement of, 83–85
    as chaff dung, 84–85
    C:N ratio of, 84–86
    combined with cover crops, 52–53
    as foundation of friability, 113
    humus production from, 75–76, 106–107
    in livestock feed production, 106
    in livestock-free agriculture, 106–107, 109
    nutrient content of, 53
    rotted, 84–85, 113
    soil microorganisms' feeding behavior in, 80–81
    as soil microorganisms' nutrient source, 44, 45, 74, 75
    straw, 85
Manure mist, 47
Manure spreaders, 84, 85
Meal worms, 78–79
Microerosion, 20–23, 24–25
Microfauna. *See also* Biotic communities
    feeding behavior of, 80–81
    role in humus formation, 76
    role in plant decomposition, 76, 77
Microflora. *See also* Biotic communities
    role in humus formation, 76
Microorganisms, in the soil
    in cultivated *versus* uncultivated fields, 73–74
    effect on silting, 13–14
    effect on topsoil quality, 1–2
    habitat of, 2
    humus formation by. *See* Humus management
    lifespans of, 12, 15
    nutrients for, 15-16, 33. *See also* Humus management
    animal-based, 77–79
    root matter, 17–19, 44–46
    oxygen requirements of, 120

role in biological tillage production, 9–19

role in humus formation. *See* Humus management

weight per acre, 9–10

Moldboard, 53

Mulch layer, 83

Mustard

as cover crop, 50

nematode infestations of, 99

Myorrhizae, 94

Natural Resources Conservation Service (NRCS), 23

Nematode poisons, 99–100

Nematodes, 99–103

beneficial, 100

parasitic, 99–103

Nitrogen

contraindication as cover crop fertilizer, 47

humus content of, 87

soil content of, 85, 118

Nitrogen surge, 95–96

"Nutrient piles," 87

Oats, soil fatigue resistance of, 95

Oxygen, 4–5, 120

pH, of soil, 114–115

Phosphate fertilizers, 115–117

Phosphoric acid, humus content of, 87

Phosphorus fertilizers, 46–47

Plants

decomposition products of, 76, 77–78, 81, 98–99

role in biological tillage, 15–16

Plowing, 119–134

"cold," 127–128

effect on soil microorganisms, 93

as friability improvement method, 43–44, 53–54

as friability loss cause, 56

functions of, 119

mistakes in, 120–124

proper methods of, 123–127

timing of, 127–128

Plow pan, 120–121, 127

Plows

animal-drawn, 127

body of, 127

stubble, 48, 128–129

two-layer, 123–124

Pores (air channels), in soil

effect of water saturation on, 21

function of, 2

in soil deterioration, 4

Potash fertilizers, 46–47, 114, 115

Potatoes, effect of soil compaction on, 39, 40
Proteolysis products, 95–96

Quicklime, 115

Rain capacity, 59–60
   irrigation systems and, 63
Rainfall
   effect on compacted topsoil, 36–37
   effect on soil crumb structure, 13–14
   effect on soil structural breakdown, 33
Rapeseed
   as cover crop, 130
      dew condensation in, 65
      disk harrowing of, 130
      effect on soil micro-organisms, 74, 75
      in livestock-free agriculture, 107–108, 109–110
      for nematode control, 101–103
   effect of soil compaction on, 39, 40
   as humus material, 78
   Liho, 99
Remedies, for unhealthy soil, 42, 43–56
Resistance factors, in soil, 96–98

Rhizosphere
   definition of, 92
   parasitic, 93, 94, 95–96, 98
   symbiotic, 92, 93–94
Roller harrows, 128
Root crops
   effect on soil friability, 105
   organic soil crumb layer for, 83
   turning of, 122
Root hairs, bacterial breakdown of, 44, 45
Root matter
   cover cropping of, 46–56
   effect on soil pH, 115
   insufficient, 44–45
   as soil organisms' nutrient source, 17–19, 44–46
Roots. *See also* Rhizosphere
   capillary water transport in, 57–58
   effect of phosphate fertilization on, 115–117
   effect of soil compaction on, 34–35
   fungal growth in, 94–95, 97
   as microorganisms' nutrient source, 44, 45
   optimal depth of, 58–59
   role in soil friability, 43–44
   as soil condition indicators, 37

as soil structure indicators, 40
in spade test, 68
Rye
  interplanted with sweet peas, 35
  phosphate fertilization of, 115–117
  soil fatigue resistance of, 95

Scheffler, F., 76
Sekera, F., 10, 77, 87, 94–95, 99, 120
Silting, 6, 7
  effect of soil microorganisms on, 13–14
  process of, 21, 27–29
  protection against, 16
  relationship to soil structural stability, 6
  of winter-plowed furrows, 127–128
Slopes, as soil erosion risk, 24
Sodium nitrate fertilizers, 114
Soil climate, 15
Soil fatigue, 91–123
  causes of, 91–103
    fungal growth, 94–95, 97
    nematodes, 99–103
  definition of, 91
Soil quality
  importance of, 19
  relationship to soil structure, 1–2
Soil sampling. *See also* Spade test
  for humus testing, 88–90
Soil structure. *See also* Aggregate structure, of soil; Crumb structure, of soil
  deterioration of, 4
    microerosion-related, 20–23
  relationship to soil quality, 1–2
  soil-compacting process of, 16–17
  soil-mellowing process of, 16–17
Soil types, relationship to water storage capacity, 59–60
Spade test, 39–41, 54, 66–71
  equipment for, 66–67
  goal of, 67
  multiple, 70
  of rapeseed cover crops, 48
  soil structural findings in, 68–70
  for soil water absorption assessment, 62
  technique of, 67–68
Star harrows, 128
Stomata
  dew absorption in, 65
  drought-related paralysis of, 64

Straw
    conversion into humus, 106–107
    in livestock-free agriculture, 108, 198
Straw cover, 52
Stubble breaking, 128–130
Stubble plow, 48, 128–129
Subsoil
    plowing of, 123
    water absorption capacity of, 62
    water absorption rate in, 24
    water storage capacity of, 59–60
Sugar beets
    effect of erosion on, 25
    effect of soil compaction on, 30, 31, 37–39
    nematode infestations of, 99–103
    taproot system of, 37–39
    yield-soil friability relationship in, 41–42
Sweet peas, effect on soil compaction, 35
Symbiosis, 92. *See also* Biocenosis; Rhizosphere, symbiotic

Taproot plants, 37–39
Tillage. *See also* Biological tillage
    for non-friable soil improvement, 53–54
    two-layer, 53
Tilth
    clover-related, 35
    "good," 1, 14
    importance of, 17–19
Topsoil
    consistency of, 7, 8
    depth of, 119–123
    factors affecting, 1–2
    root spread in, 122
    structural breakdown of
        in cereal crops, 30–35
        cross-sectional diagram of, 27–29
        effect of clover seeds on, 33–35
        progression of, 27–35
        timing of onset of, 30–35
        waterlogged condition of, 36–37
Trace elements, humus content of, 87
Tractors, wheel pressure of, 131–134

Vageler, P., 10
von Rümker, K., 19

Water. *See also* Rainfall
    effect on topsoil aggregate structure, 9

as microerosion cause, 10–11, 20–23
Water (rain) capacity, 59–60
   irrigation systems and, 63
Water reservoir, soil as, 57–65
   in compacted soil, 60–62
   critical water content and, 63–64
   dew as, 65
   distribution in friable and non-friable soils, 132–133
   effect on root depth, 57–60
   fertilization and, 113
   irrigation systems and, 62–63
   water (rain) capacity of, 59–60, 63
Water resistance, effect on topsoil quality, 1–2
Water storage, in soil, 2
Wheat. *See also* Winter wheat
   in livestock-free agriculture, 109–110

Wheat stubble, soil microorganisms in, 74, 75
White clover, as cover crop, 51
Wild rye, in compacted soil, 121
Wind, as erosion cause, 26
Winter. A., 94
Winter wheat
   nematode control in, 101–103
   phosphate fertilization of, 116, 117
   yield-soil friability relationship in, 41–42

Yield
   effect of soil compaction on, 30
   relationship to soil friability, 41–42
   relationship to topsoil depth, 121

About the Author

Prof. Dr. Franz Sekera (1899–1955) joined the University of Agricultural Sciences in Vienna's newly established Institute of Soil Biology in 1939, where he taught plant nutrition and was appointed full professor in 1942. He spent his career focusing on biological problems of soil fertility and tilth.

He defined tilth as "the state of soil crumbs for optimum stability." He determined that the formation of good soil structure was strongly connected with the vitality of soil microorganisms and coined the concept of the "connective tissue of soil structure." His findings gave the research on so-called "tilth" a new framework.

His legacy for the science and practice of agriculture is this widely-distributed book, *Healthy Soils, Sick Soils* (originally published as *Gesund und Kranken Boden*). In this work Sekera has described cause-and-effect relationships in the soil as well as practical measures for conservation and improvement of soil tilth, presented in detail and remaining true to his motto, "always with nature and never against it!"

His wife, researcher Margareth Sekera (Dipl. Ing.), revised and expanded later editions of this classic work after the death of her husband. She was active in the International Society of Soil Science.

Acres U.S.A. — our bookstore is just the beginning!

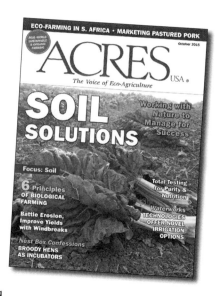

Farmers and gardeners around the world are learning to grow bountiful crops profitably — without risking their own health and destroying the fertility of the soil. *Acres U.S.A.* can show you how. If you want to be on the cutting edge of organic and sustainable growing technologies, techniques, markets, news, analysis and trends, look to *Acres U.S.A.* For over 40 years, we've been the independent voice for eco-agriculture. Each monthly issue is packed with practical, hands-on information you can put to work on your farm, bringing solutions to your most pressing problems. Get the advice consultants charge thousands for . . .

- Fertility management
- Non-chemical weed & insect control
- Specialty crops & marketing
- Grazing, composting & natural veterinary care
- Soil's link to human & animal health

To subscribe, visit us online at

www.acresusa.com

or call toll-free in the U.S. and Canada

1-800-355-5313

Outside U.S. & Canada call 512-892-4400
fax 512-892-4448 • info@acresusa.com